Imaginary Abacus - Workbook

A mind math step-by-step guide to addition and subtraction using an imaginary Japanese abacus (Soroban).

Author:
Paul Green

First published 2017

ISBN-13: 978-1542357227
ISBN-10: 1542357225

PREFACE

This book is a workbook which is to be used with the instruction book called ‘Imaginary Abacus - Instruction book’ (ISBN: 978-1978043497).

WORKBOOK WORK - Progress sheet

INSTRUCTION BOOK	Part 1		WORKBOOK	Page 8	
			WORKBOOK	Page 9	
			WORKBOOK	Page 10	
			WORKBOOK	Page 11	
			WORKBOOK	Page 12	
			WORKBOOK	Page 13	
INSTRUCTION BOOK	Part 2		WORKBOOK	Page 14	
			WORKBOOK	Page 15	
			WORKBOOK	Page 16	
			WORKBOOK	Page 17	
INSTRUCTION BOOK	Part 3		WORKBOOK	Page 18	
			WORKBOOK	Page 19	
			WORKBOOK	Page 20	
			WORKBOOK	Page 21	
			WORKBOOK	Page 22	
INSTRUCTION BOOK	Part 4		WORKBOOK	Page 23	
			WORKBOOK	Page 24	
			WORKBOOK	Page 25	
			WORKBOOK	Page 26	
			WORKBOOK	Page 27	
			WORKBOOK	Page 28	
			WORKBOOK	Page 29	
INSTRUCTION BOOK	Part 5		WORKBOOK	Page 30	
			WORKBOOK	Page 31	
			WORKBOOK	Page 32	

WORKBOOK WORK - Progress sheet

			WORKBOOK	Page 33	
			WORKBOOK	Page 34	
			WORKBOOK	Page 35	
INSTRUCTION BOOK	Part 6		WORKBOOK	Page 36	
			WORKBOOK	Page 37	
			WORKBOOK	Page 38	
			WORKBOOK	Page 39	
			WORKBOOK	Page 40	
			WORKBOOK	Page 41	
			WORKBOOK	Page 42	
INSTRUCTION BOOK	Part 7		WORKBOOK	Page 43	
			WORKBOOK	Page 44	
			WORKBOOK	Page 45	
			WORKBOOK	Page 46	
			WORKBOOK	Page 47	
			WORKBOOK	Page 48	
INSTRUCTION BOOK	Part 8		WORKBOOK	Page 49	
			WORKBOOK	Page 50	
			WORKBOOK	Page 51	
INSTRUCTION BOOK	Part 9		WORKBOOK	Page 52	
			WORKBOOK	Page 53	
			WORKBOOK	Page 54	
INSTRUCTION BOOK	Part 10		WORKBOOK	Page 55	
			WORKBOOK	Page 56	
			WORKBOOK	Page 57	

CONTENTS

HOW TO FOLLOW THIS TRAINING COURSE

THE BEST WAY TO PROCEED

1. Read through the instructions in the book 'Imaginary Abacus - Instruction book' (ISBN: 978-1978043497), at your own pace, until you see this note:

> Time to use the **workbook!** Go to workbook **page** 8.

2. Open this workbook at the page stated in the instruction book and follow the work given until you see this note in this book:

> Time to use the **instruction book!** Go to instruction book **page** 20.

3. Go back to your instruction book and continue in this way.

4. Keep a track of your progress by ticking the appropriate box next to the instruction work and your workbook work (see your instruction book).

INSTRUCTION BOOK	Part 1	✓	WORKBOOK	Page 8	✓
			WORKBOOK	Page 9	✓
			WORKBOOK	Page 10	✓
			WORKBOOK	Page 11	✓
			WORKBOOK	Page 12	✓
			WORKBOOK	Page 13	✓
INSTRUCTION BOOK	Part 2		WORKBOOK	Page 14	
			WORKBOOK	Page 15	

WORKBOOK WORK - 1

(Answers to workbook work 1 are on pages 74 to 76)

Draw the beads on the empty abacus to represent the number given.

Examples:

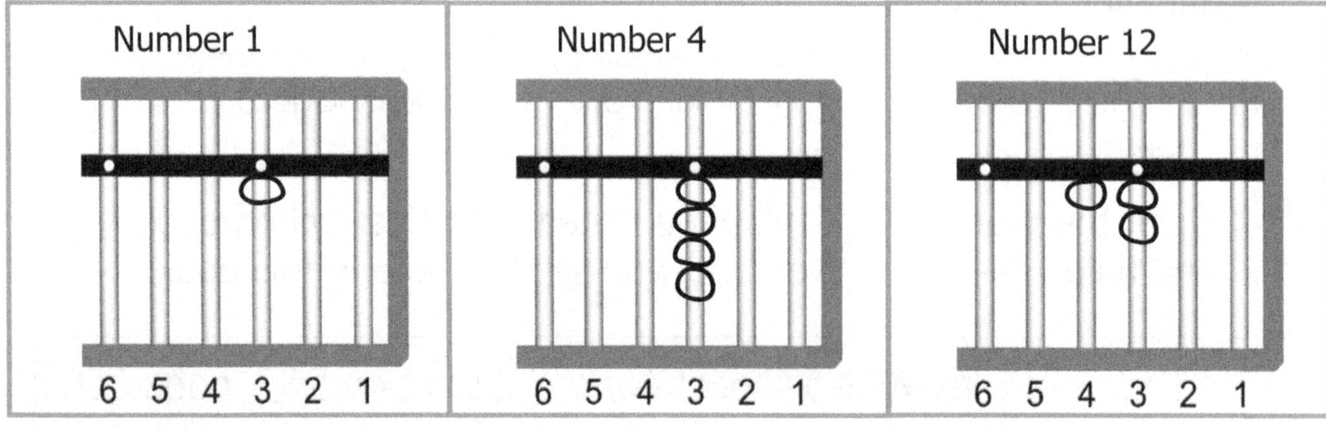

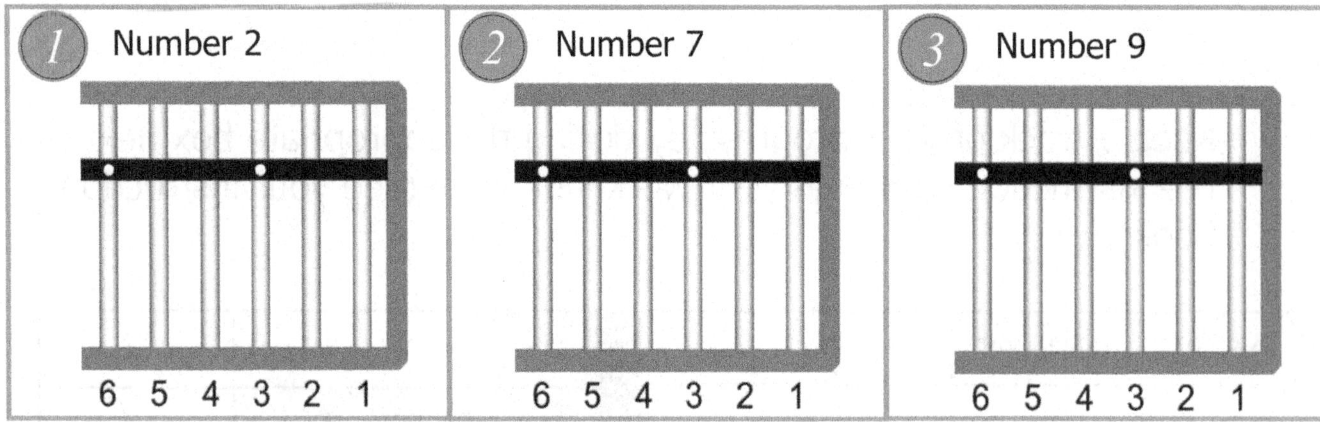

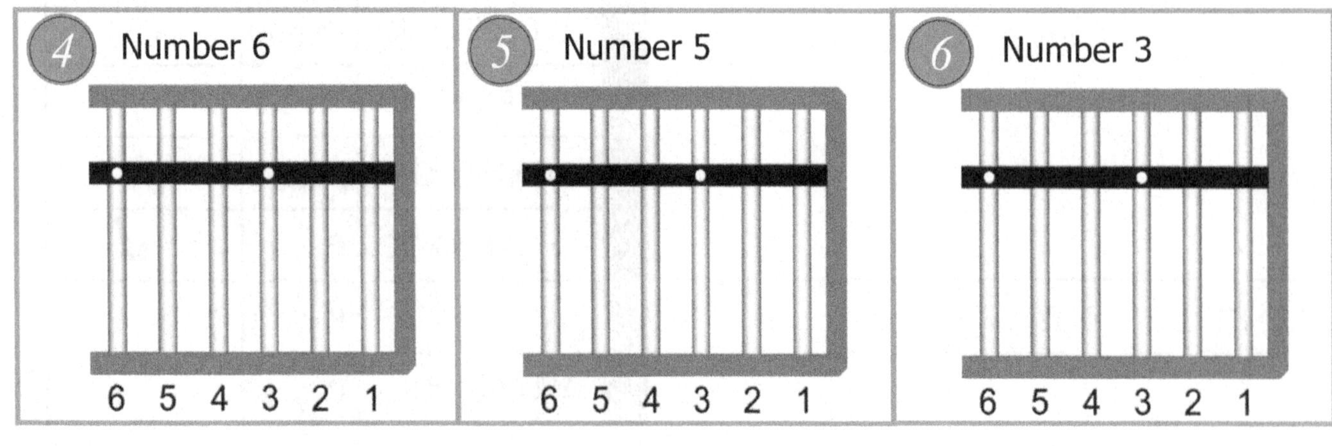

Pencil

WORKBOOK WORK - 1

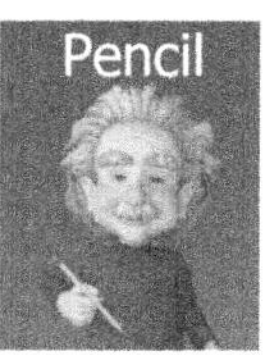

7 Number 8

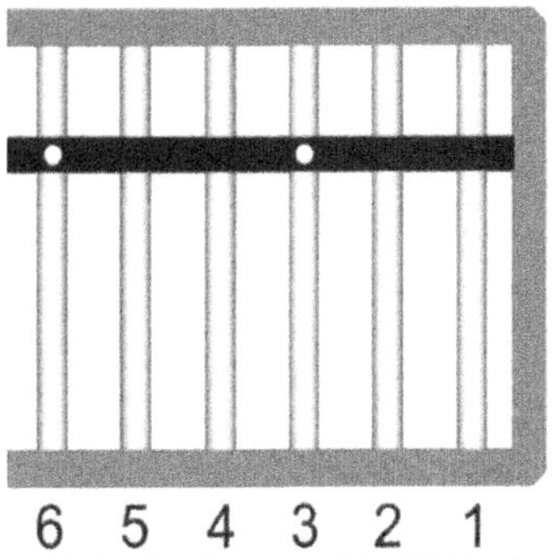
6 5 4 3 2 1

8 Number 18

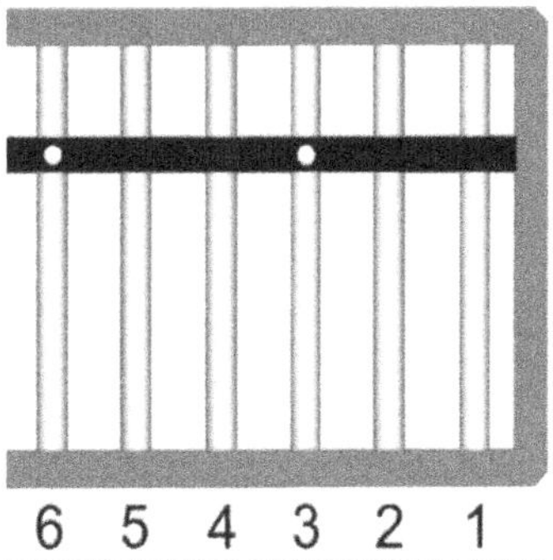
6 5 4 3 2 1

9 Number 16

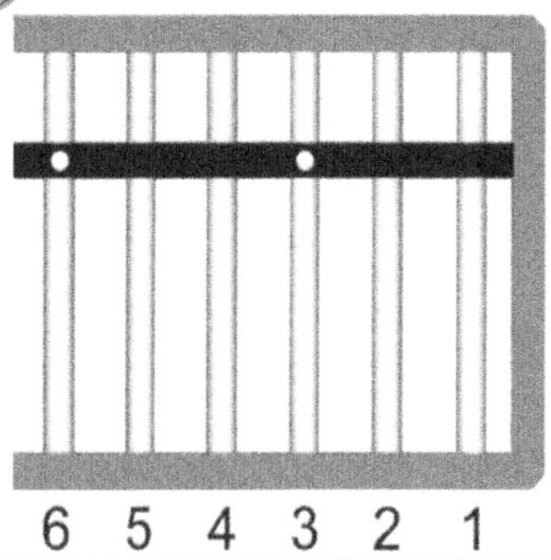
6 5 4 3 2 1

10 Number 14

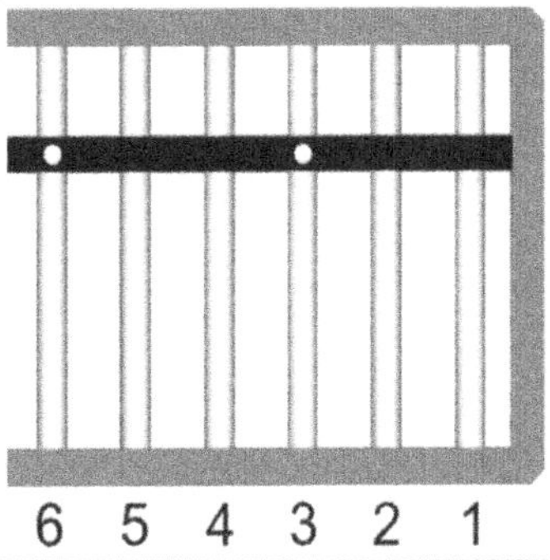
6 5 4 3 2 1

11 Number 13

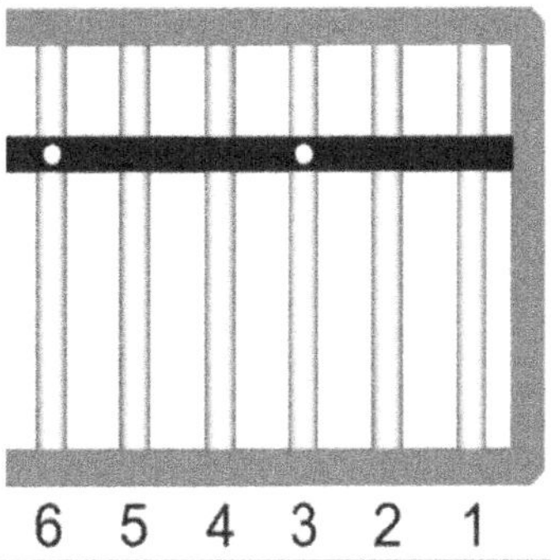
6 5 4 3 2 1

12 Number 11

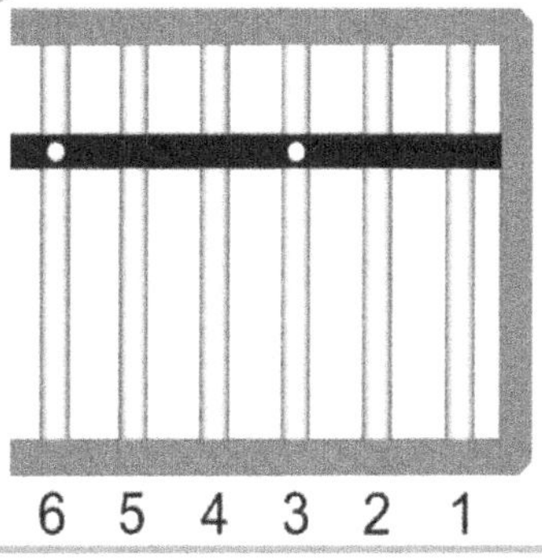
6 5 4 3 2 1

13 Number 19

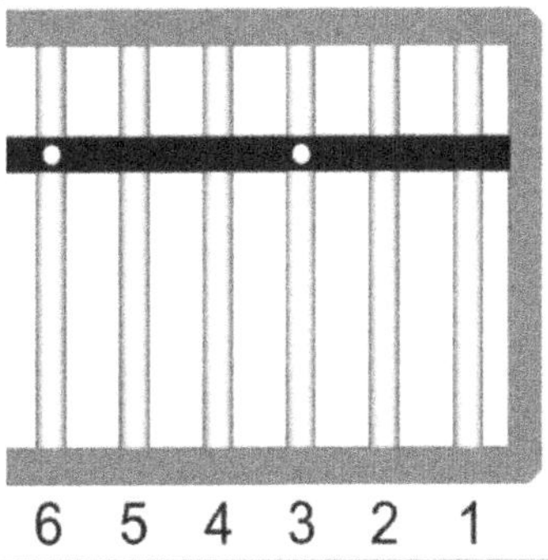
6 5 4 3 2 1

14 Number 15

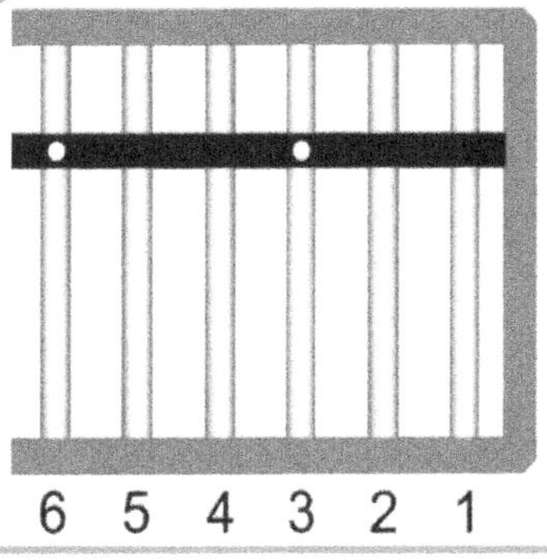
6 5 4 3 2 1

15 Number 10

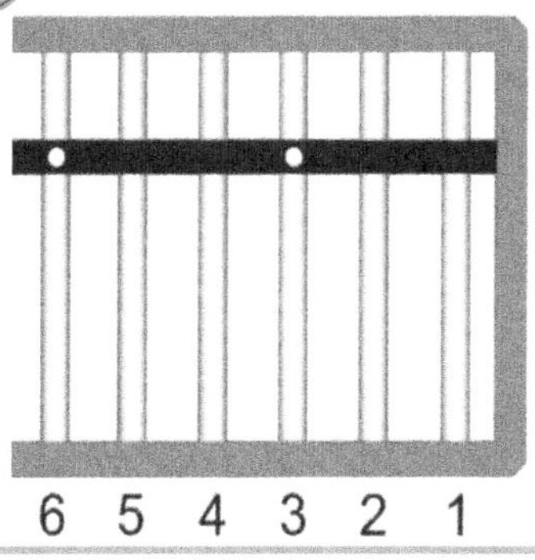
6 5 4 3 2 1

16 Number 17

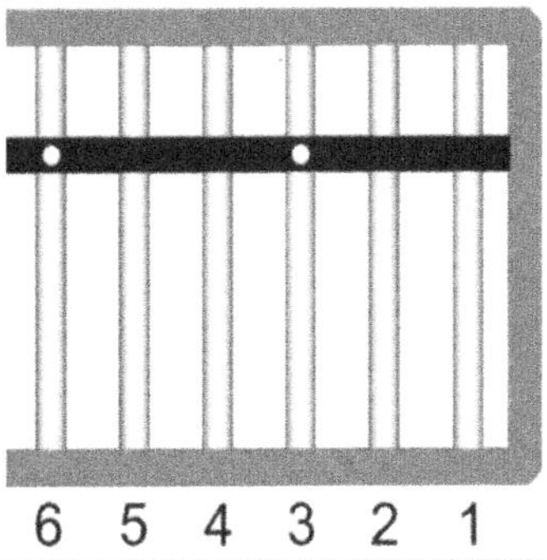
6 5 4 3 2 1

17 Number 20

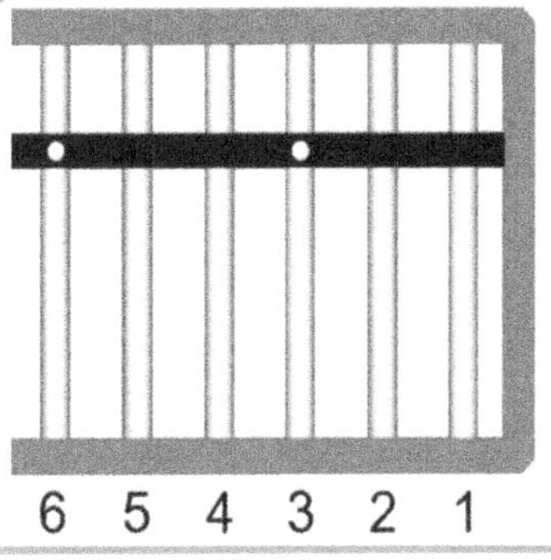
6 5 4 3 2 1

18 Number 29

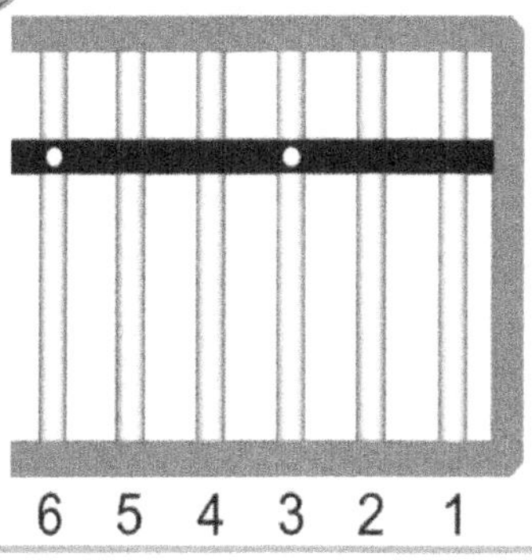
6 5 4 3 2 1

WORKBOOK WORK - 1

Pencil

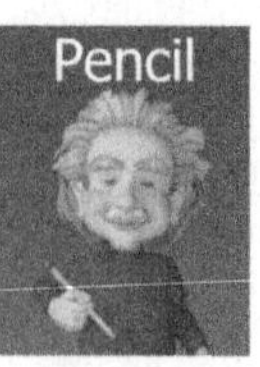

19 Number 26

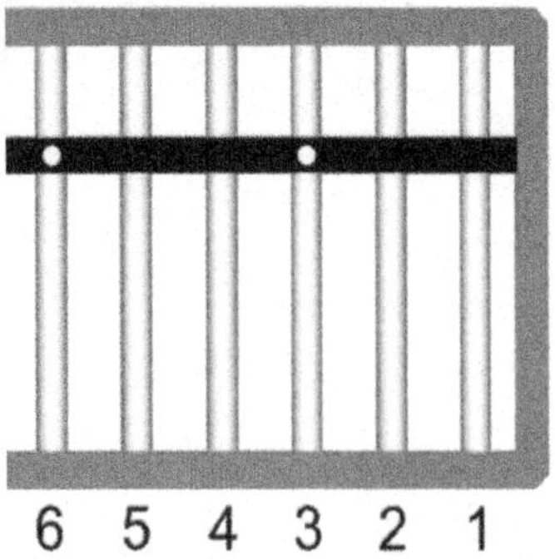
6 5 4 3 2 1

20 Number 38

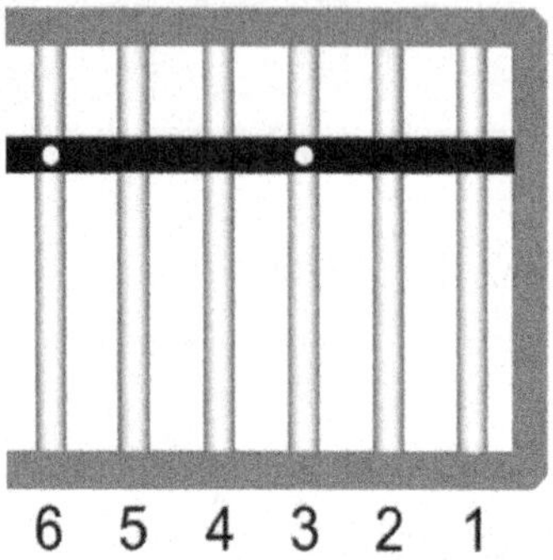
6 5 4 3 2 1

21 Number 30

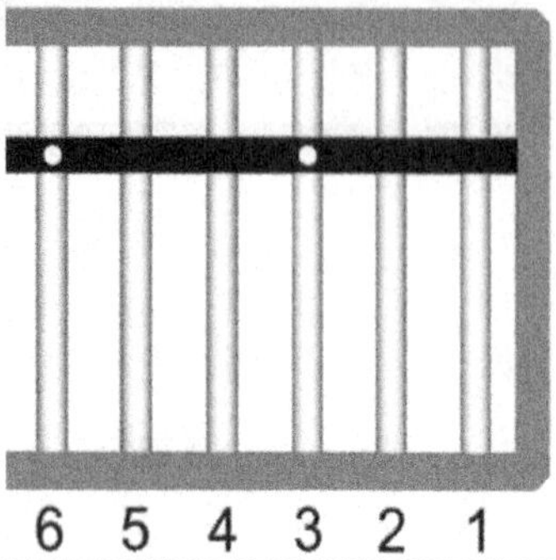
6 5 4 3 2 1

22 Number 41

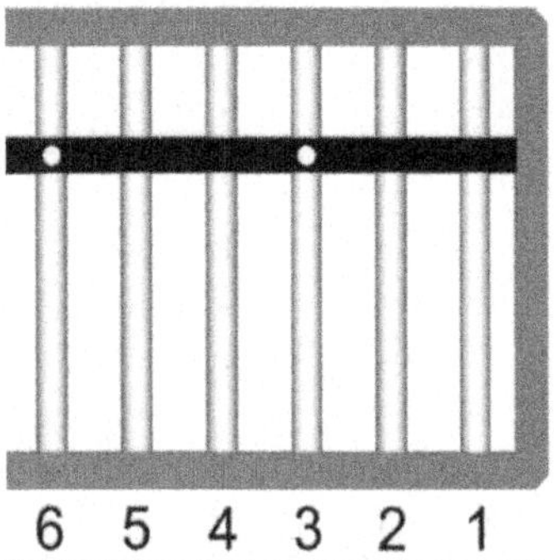
6 5 4 3 2 1

23 Number 21

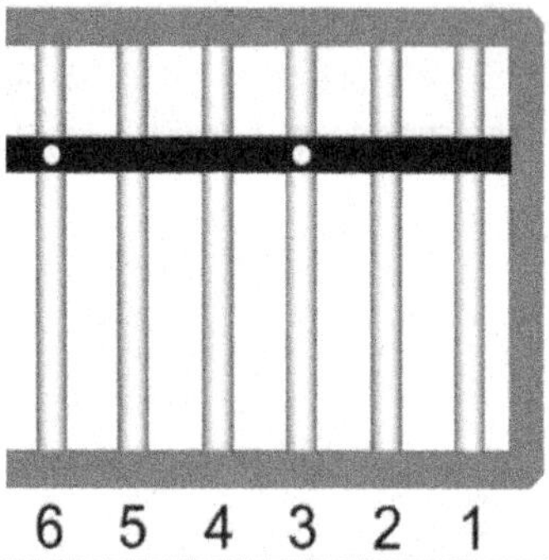
6 5 4 3 2 1

24 Number 11

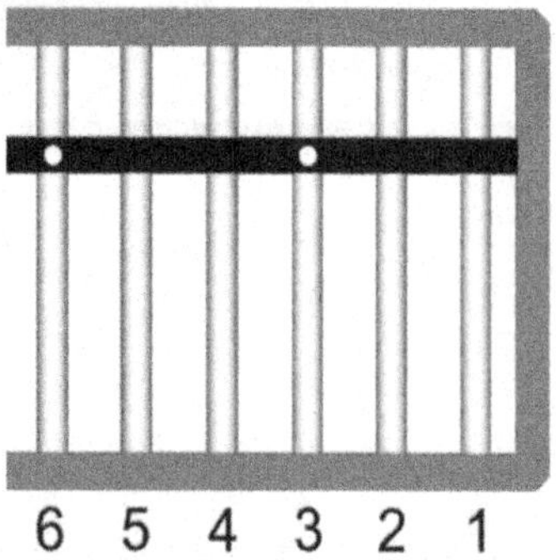
6 5 4 3 2 1

25 Number 43

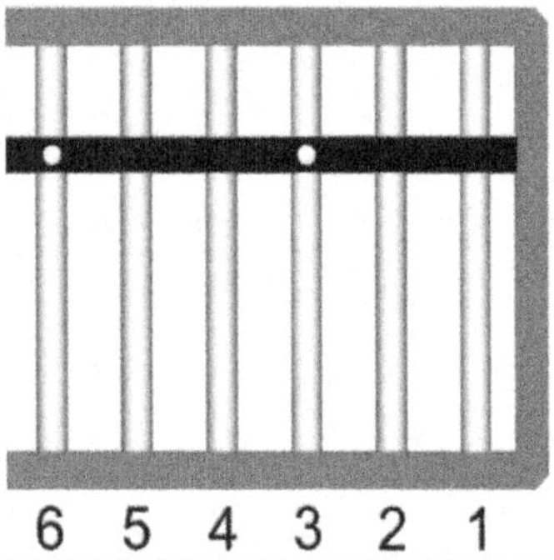
6 5 4 3 2 1

26 Number 50

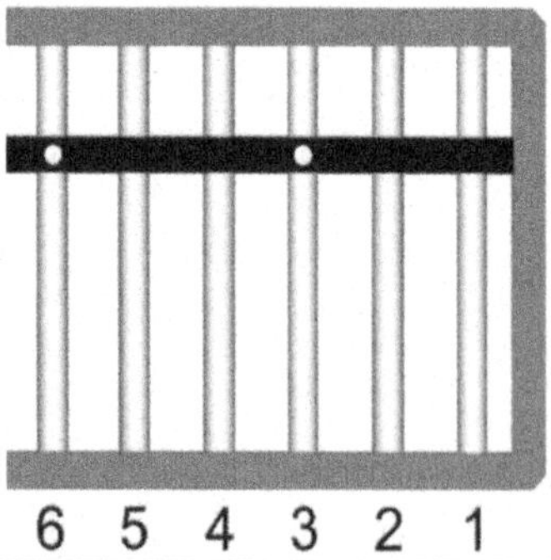
6 5 4 3 2 1

27 Number 48

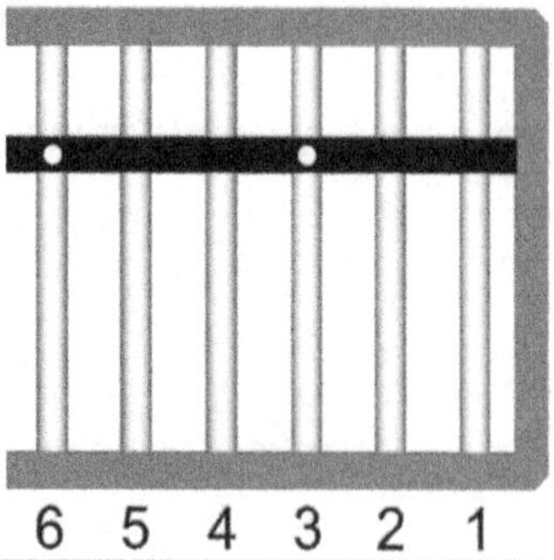
6 5 4 3 2 1

28 Number 64

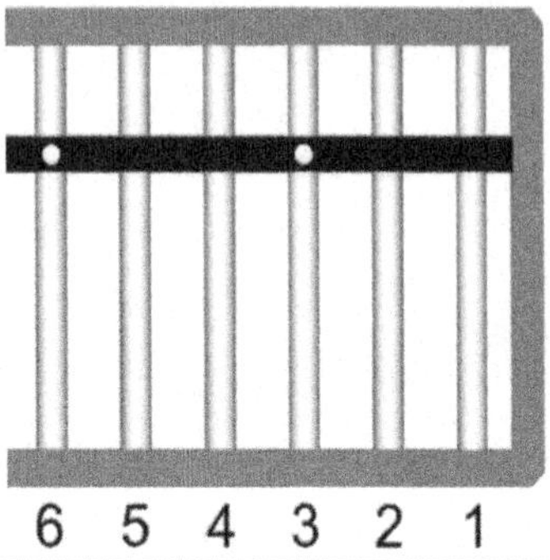
6 5 4 3 2 1

29 Number 55

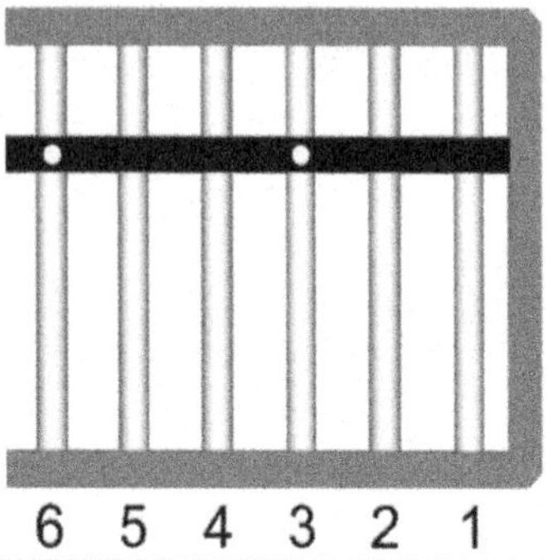
6 5 4 3 2 1

30 Number 62

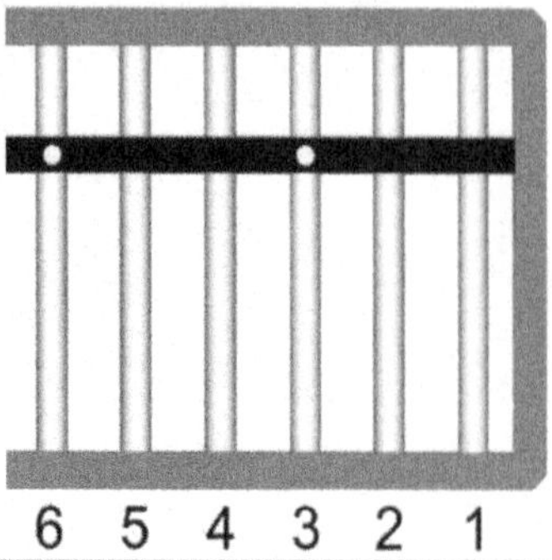
6 5 4 3 2 1

WORKBOOK WORK - 1

Pencil

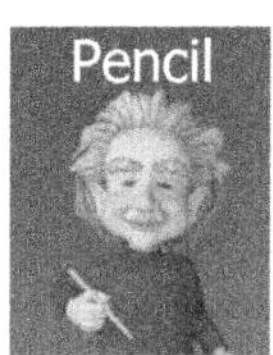

31 Number 73

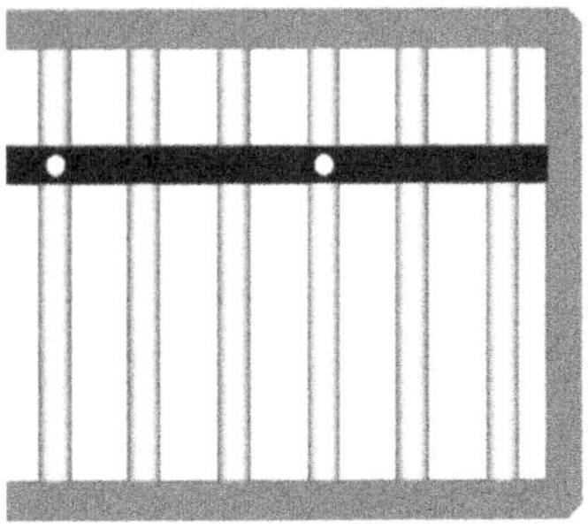
6 5 4 3 2 1

32 Number 66

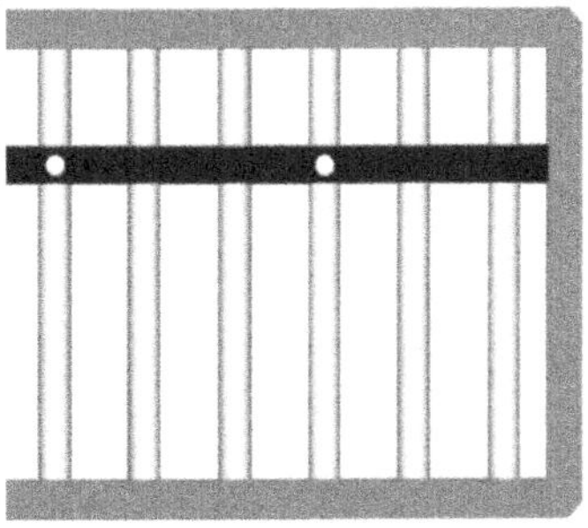
6 5 4 3 2 1

33 Number 87

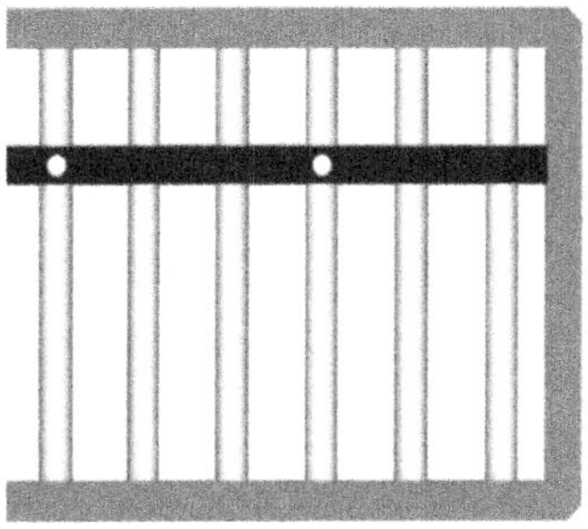
6 5 4 3 2 1

34 Number 41

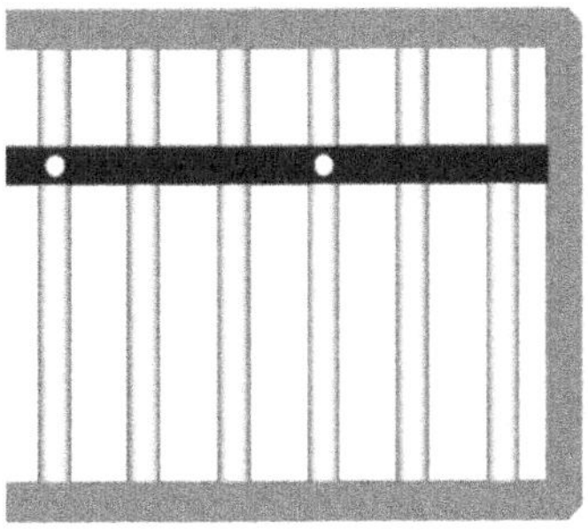
6 5 4 3 2 1

35 Number 76

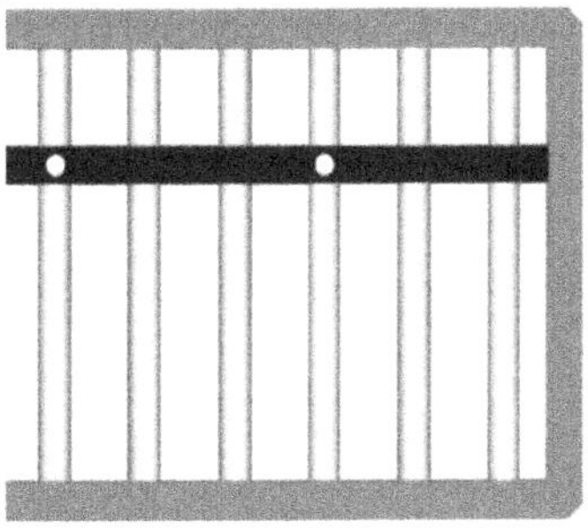
6 5 4 3 2 1

36 Number 37

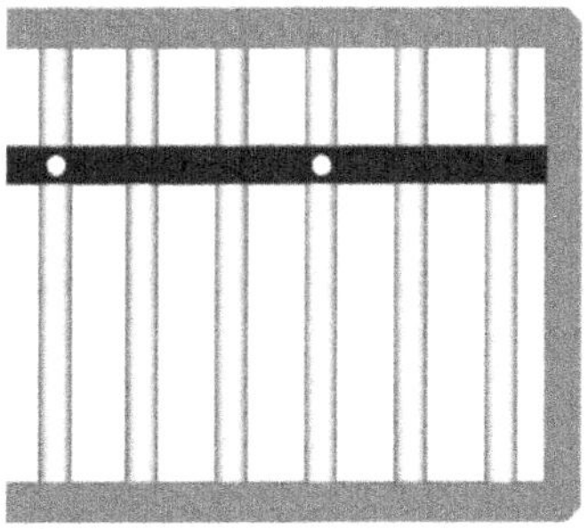
6 5 4 3 2 1

37 Number 69

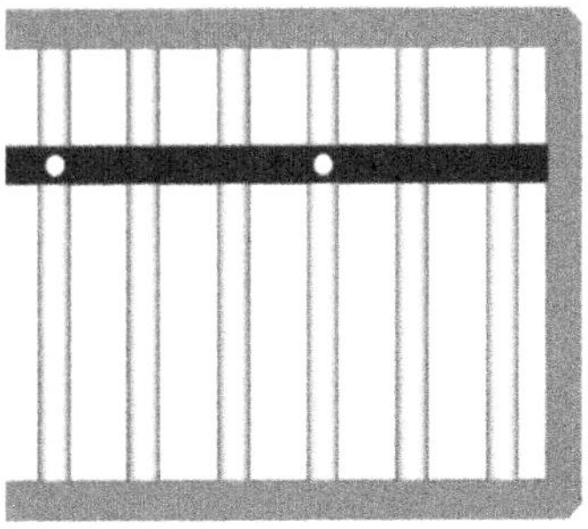
6 5 4 3 2 1

38 Number 85

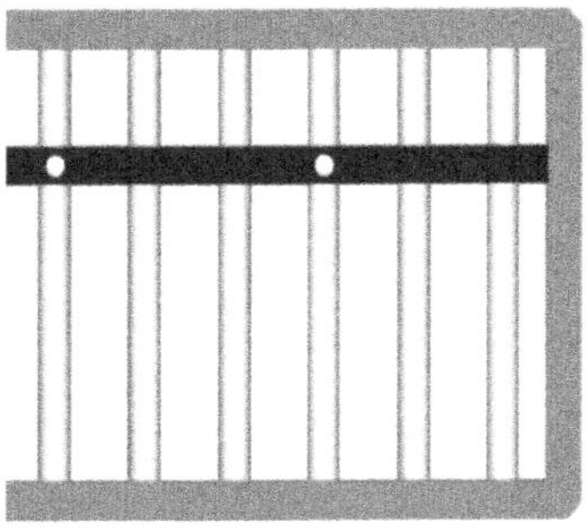
6 5 4 3 2 1

39 Number 92

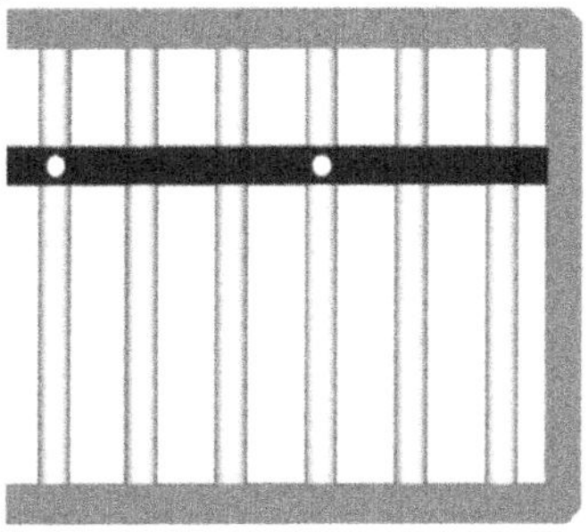
6 5 4 3 2 1

40 Number 64

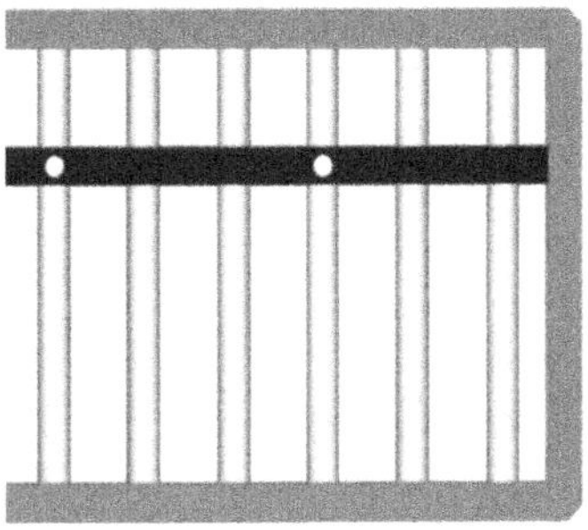
6 5 4 3 2 1

41 Number 95

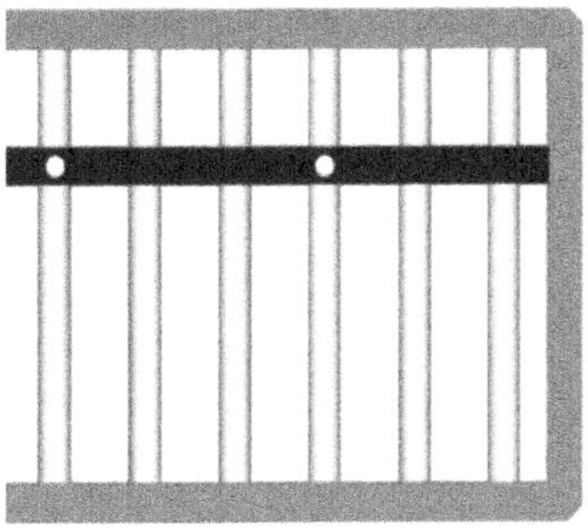
6 5 4 3 2 1

42 Number 99

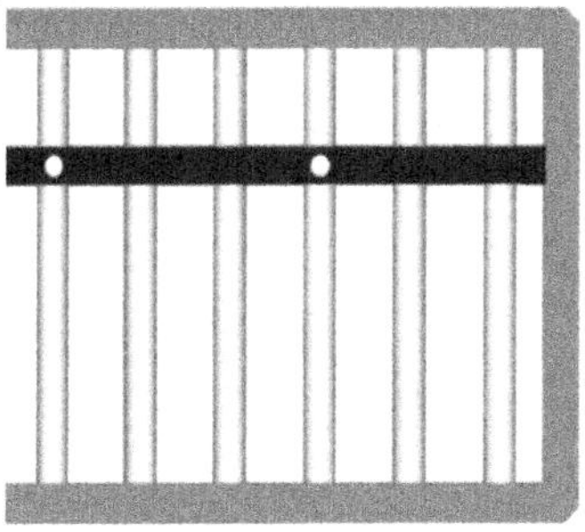
6 5 4 3 2 1

WORKBOOK WORK - 1

2 Find the correct column for the digit, by putting a circle around the column number.

Examples:

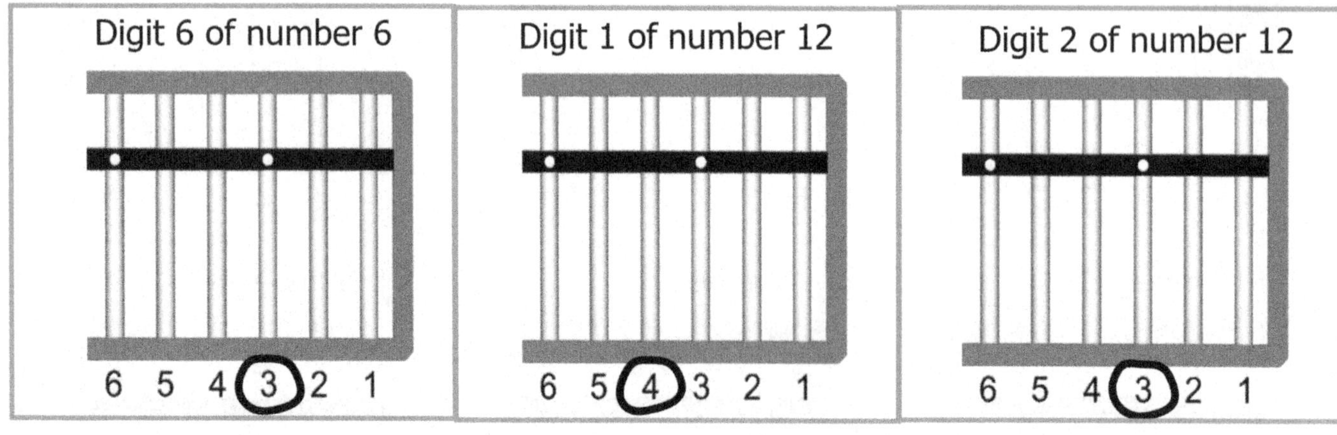

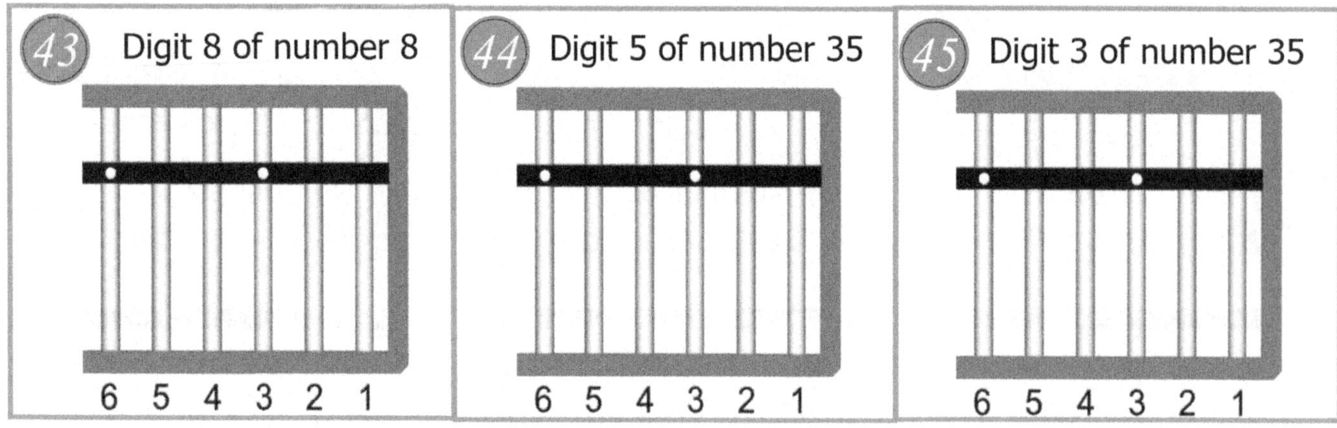

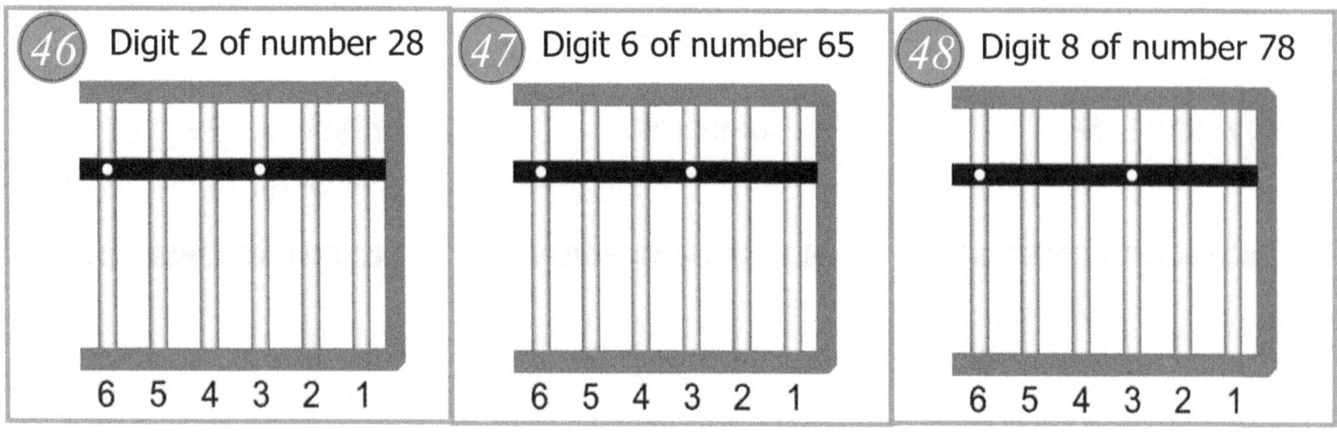

WORKBOOK WORK - 1

49 Digit 7 of number 27

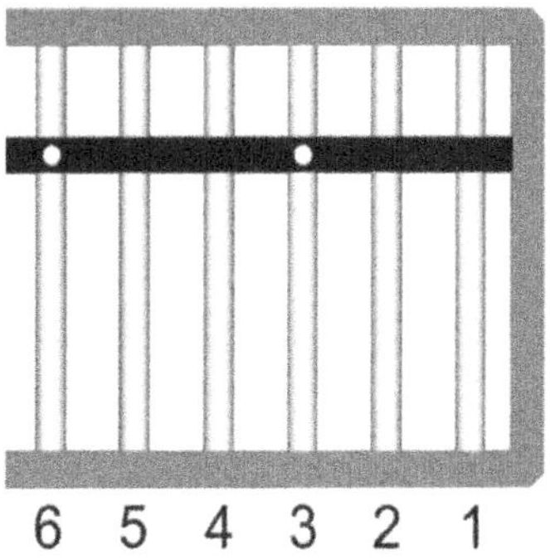

6 5 4 3 2 1

50 Digit 8 of number 48

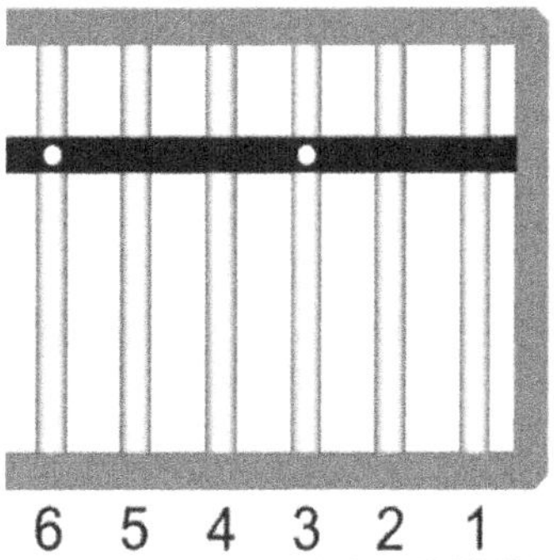

6 5 4 3 2 1

51 Digit 5 of number 56

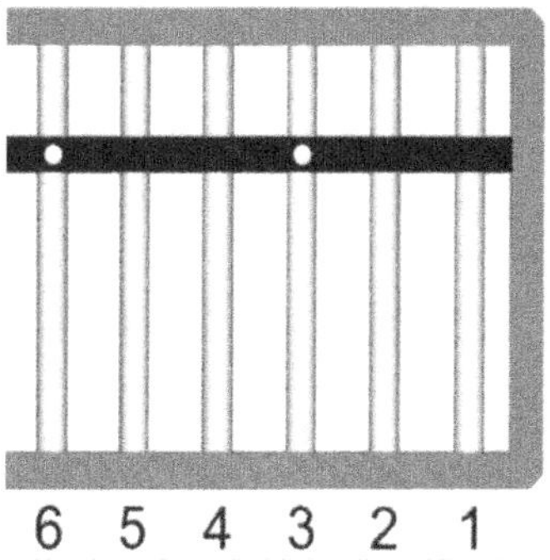

6 5 4 3 2 1

52 Digit 6 of number 61

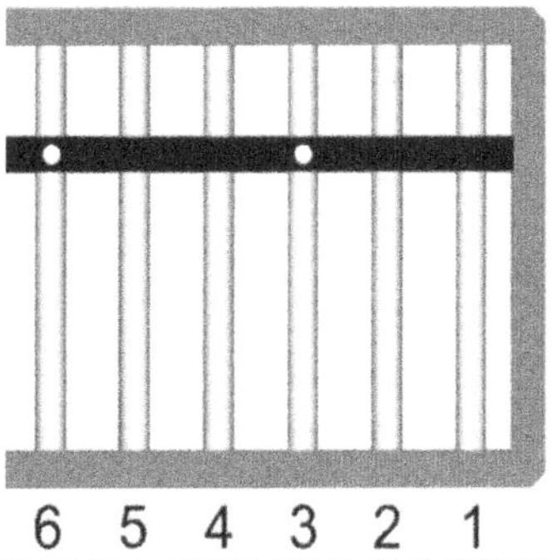

6 5 4 3 2 1

53 Digit 3 of number 83

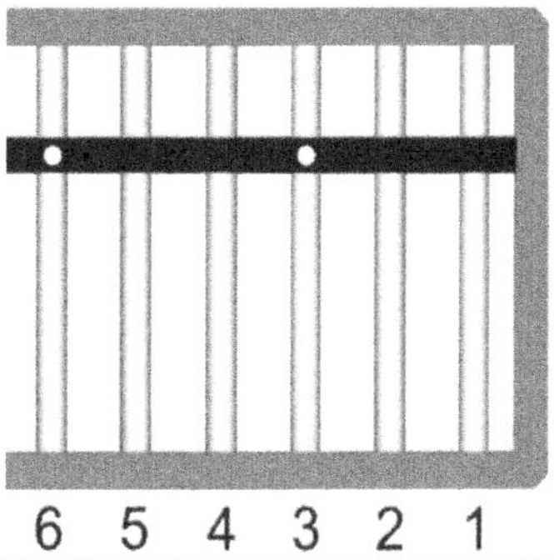

6 5 4 3 2 1

54 Digit 4 of number 42

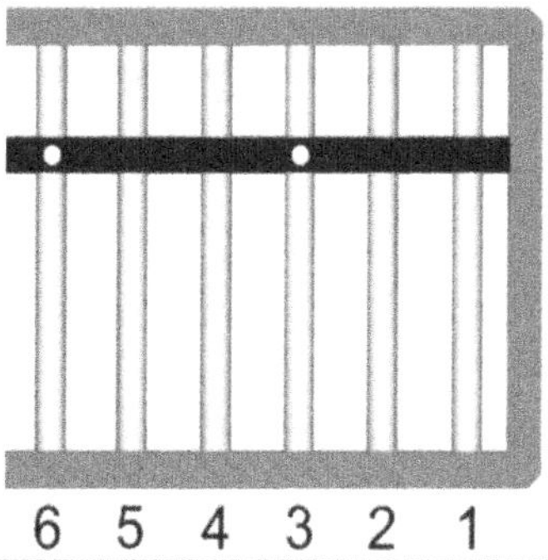

6 5 4 3 2 1

55 Digit 1 of number 21

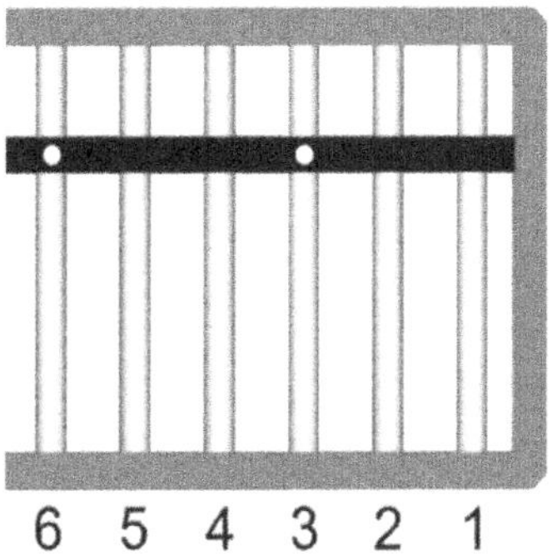

6 5 4 3 2 1

56 Digit 1 of number 15

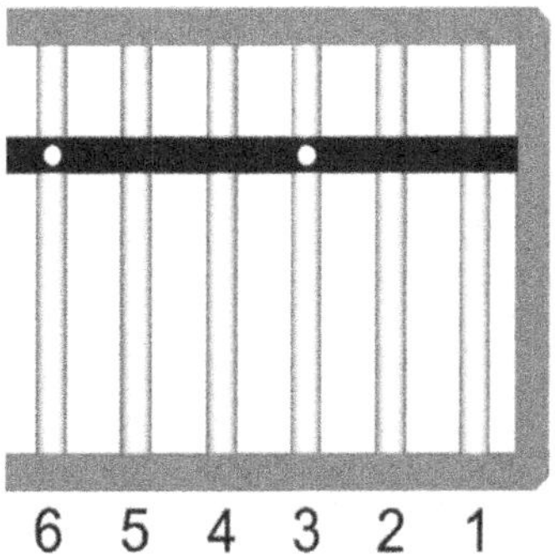

6 5 4 3 2 1

57 Digit 9 of number 94

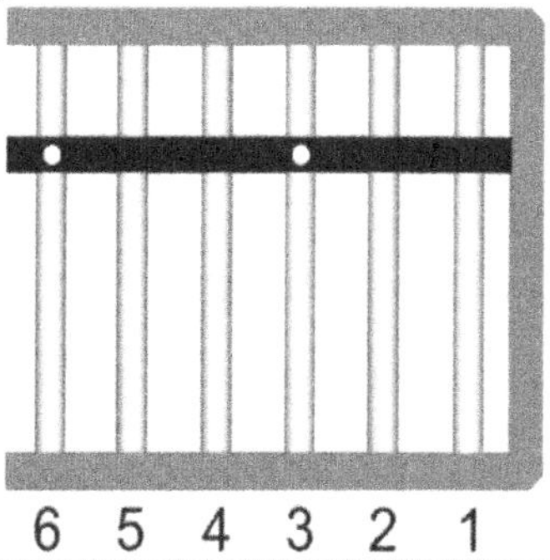

6 5 4 3 2 1

Time to use the **instruction book!** Go to instruction book **page 20.**

WORKBOOK WORK - 2

(Answers to workbook work 2 are on page 77)

1 Write down the number that is shown on the abacus.

Examples:

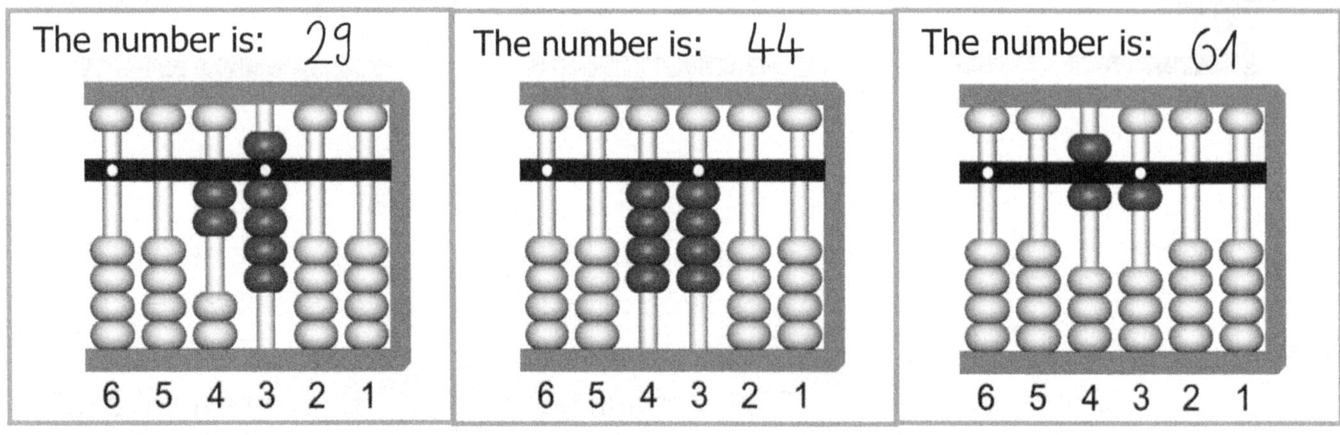

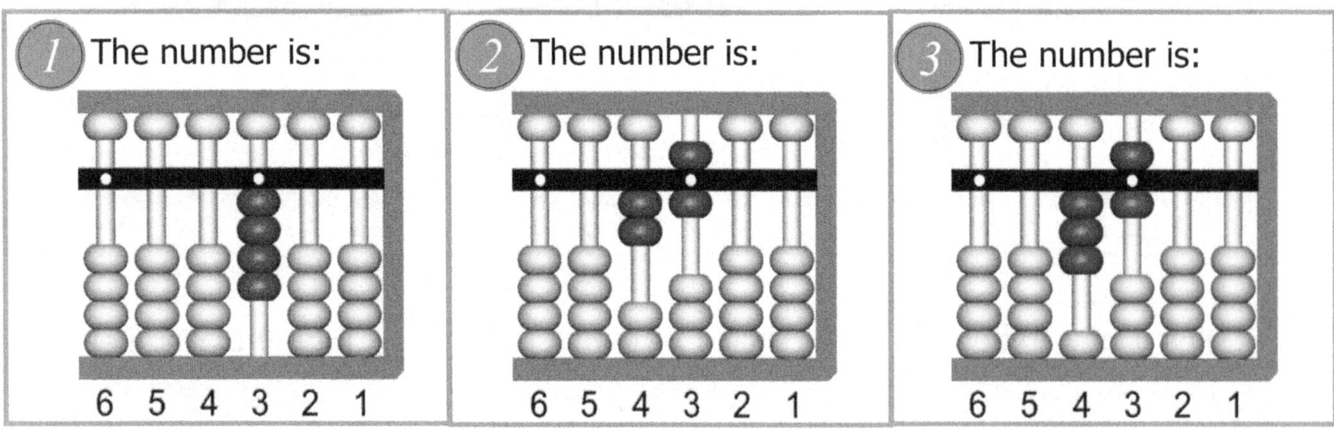

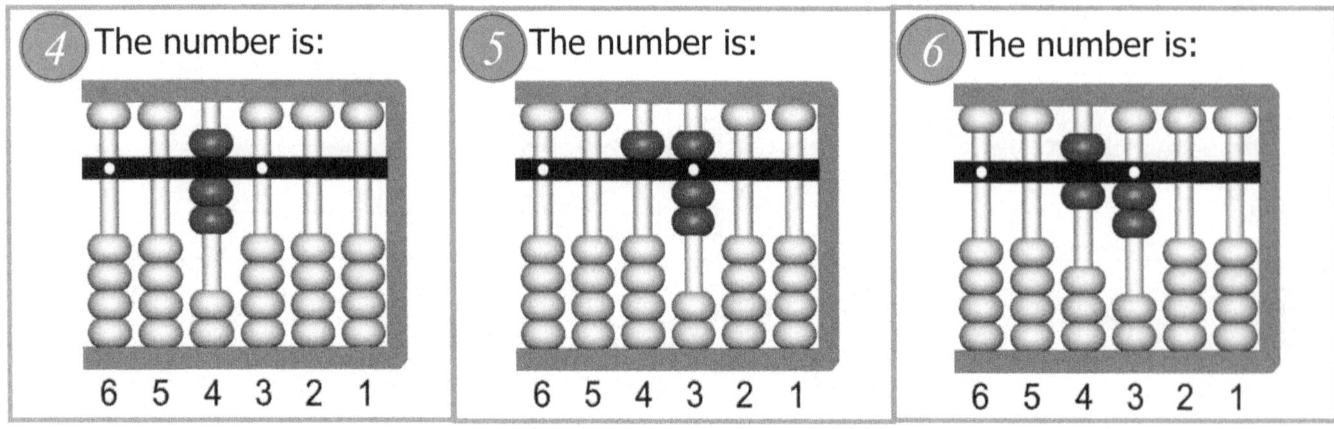

WORKBOOK WORK - 2

7 The number is:

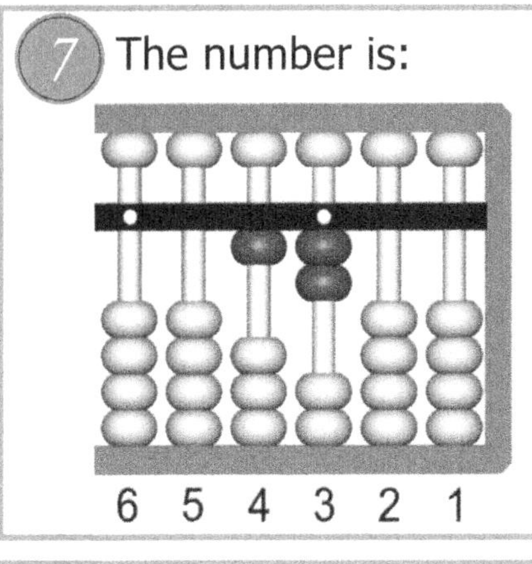

8 The number is:

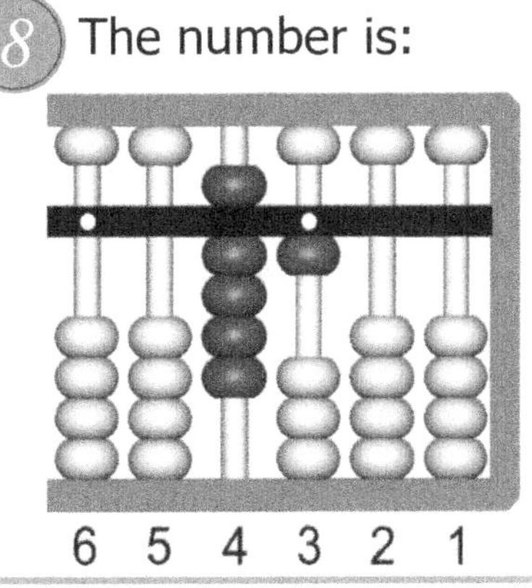

9 The number is:

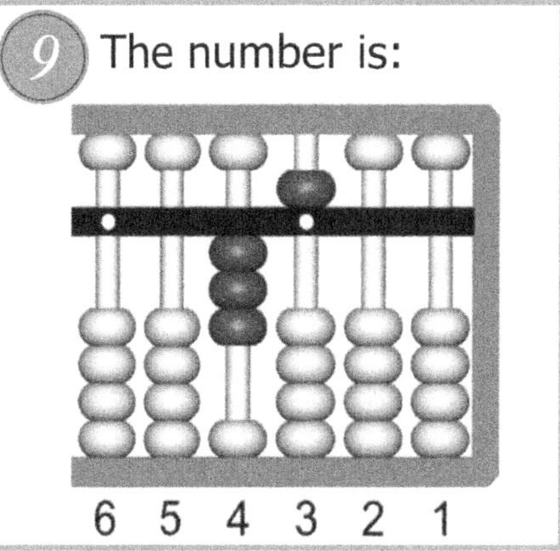

10 The number is:

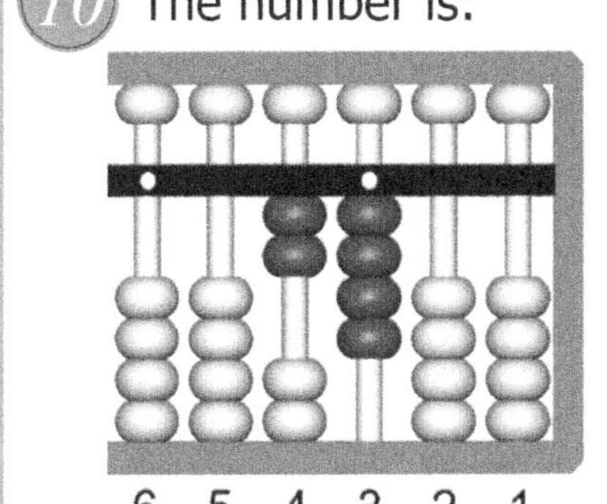

11 The number is:

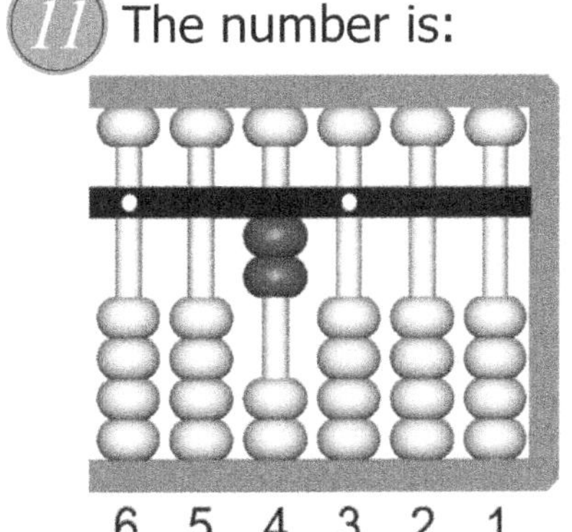

12 The number is:

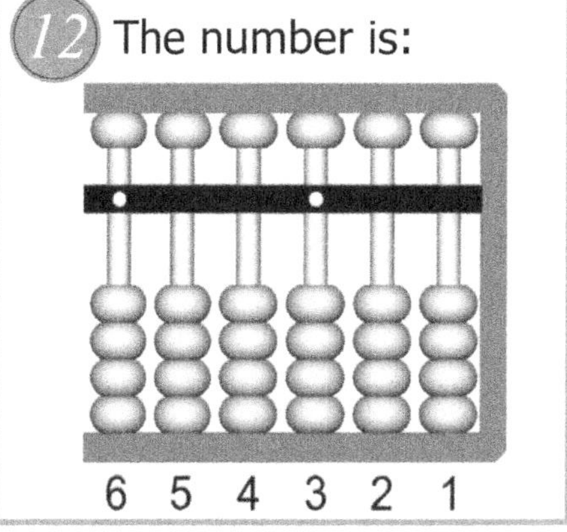

13 The number is:

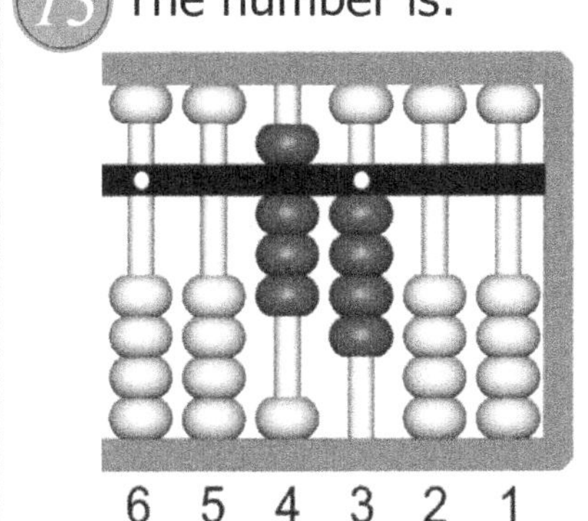

14 The number is:

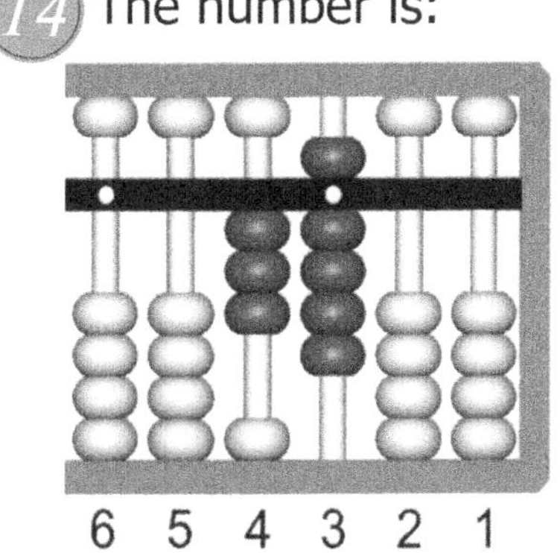

15 The number is:

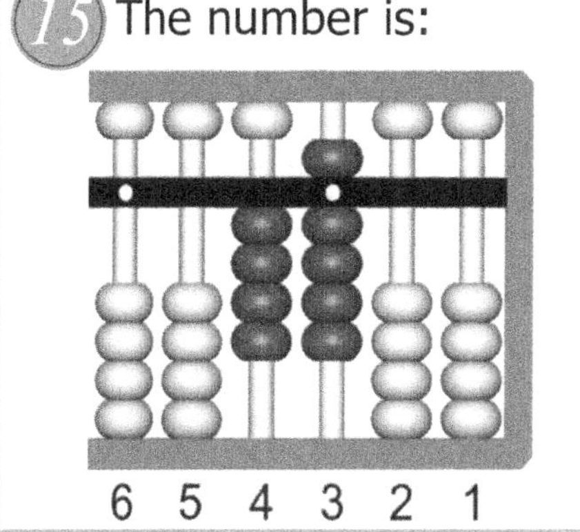

16 The number is:

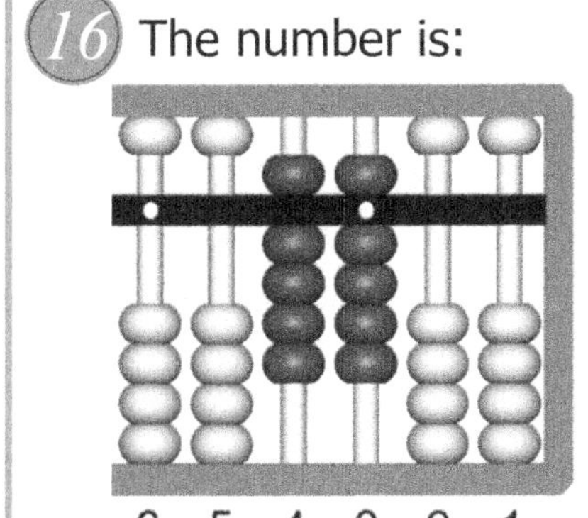

17 The number is:

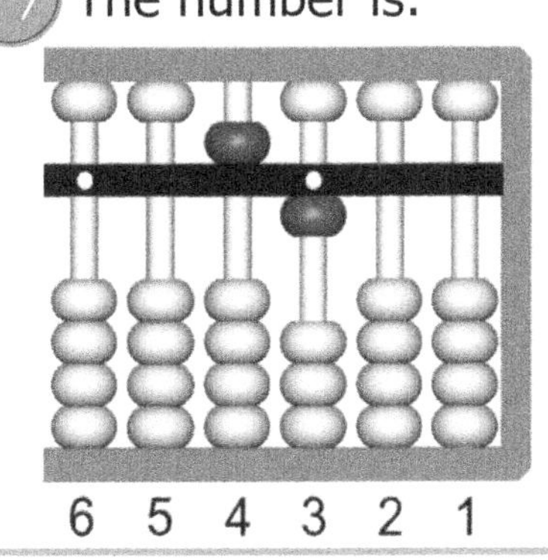

18 The number is:

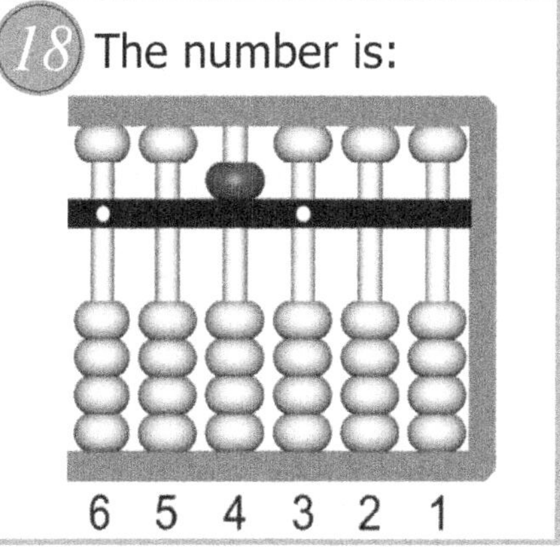

WORKBOOK WORK - 2

Abacus

2 Practise moving the beads to register these numbers.

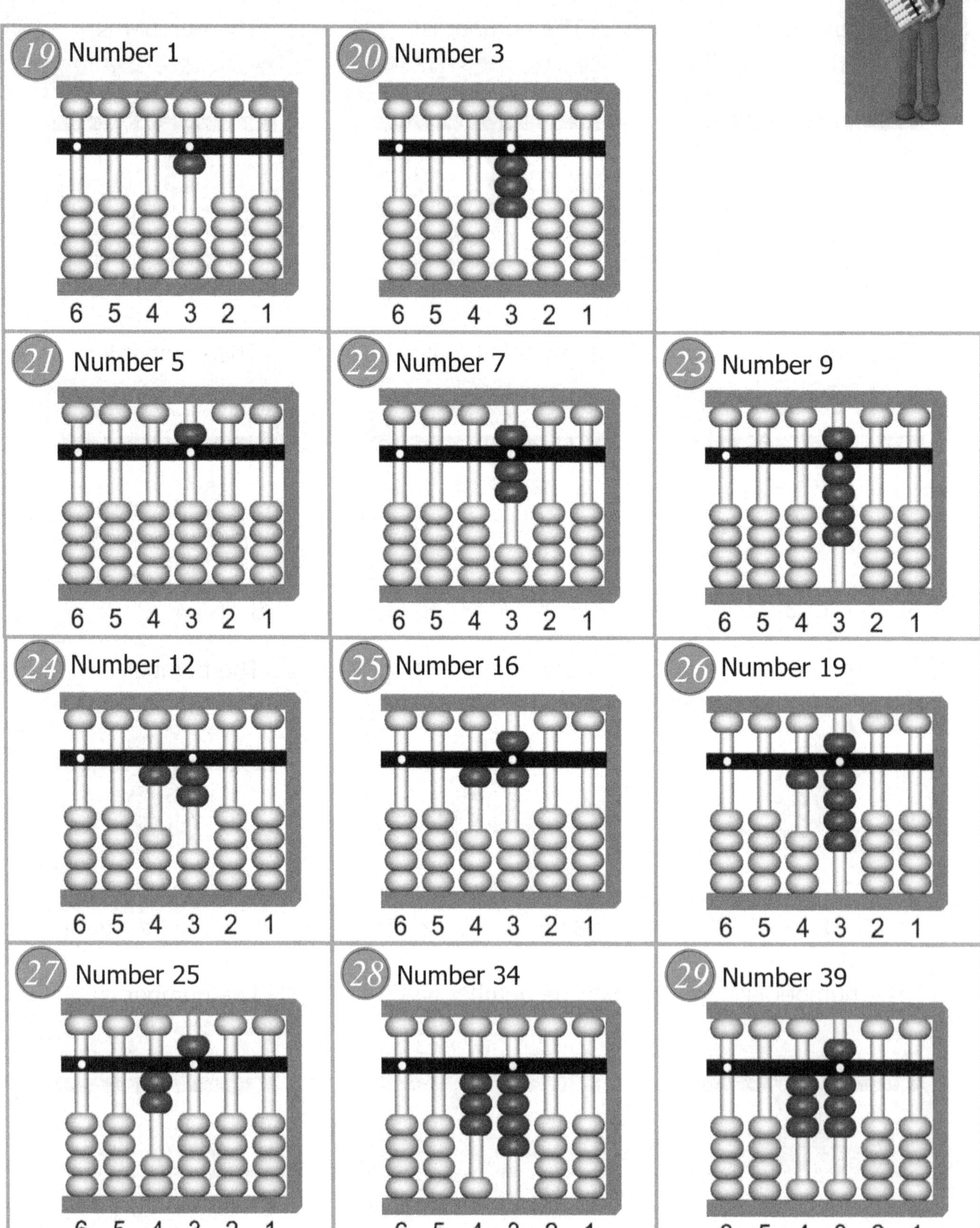

WORKBOOK WORK - 2

30 Number 43 6 5 4 3 2 1	31 Number 47 6 5 4 3 2 1	32 Number 50 6 5 4 3 2 1
33 Number 62 6 5 4 3 2 1	34 Number 68 6 5 4 3 2 1	35 Number 73 6 5 4 3 2 1
36 Number 77 6 5 4 3 2 1	37 Number 83 6 5 4 3 2 1	38 Number 89 6 5 4 3 2 1
39 Number 90 6 5 4 3 2 1	40 Number 94 6 5 4 3 2 1	41 Number 98 6 5 4 3 2 1

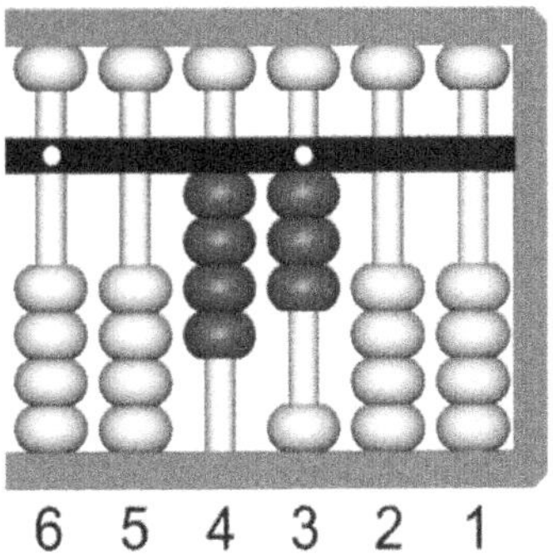

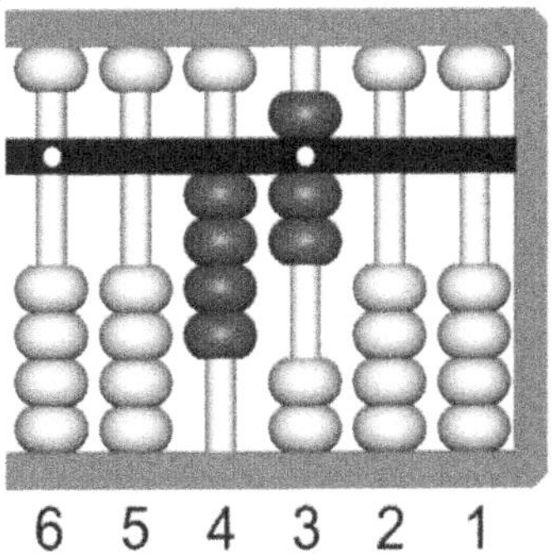

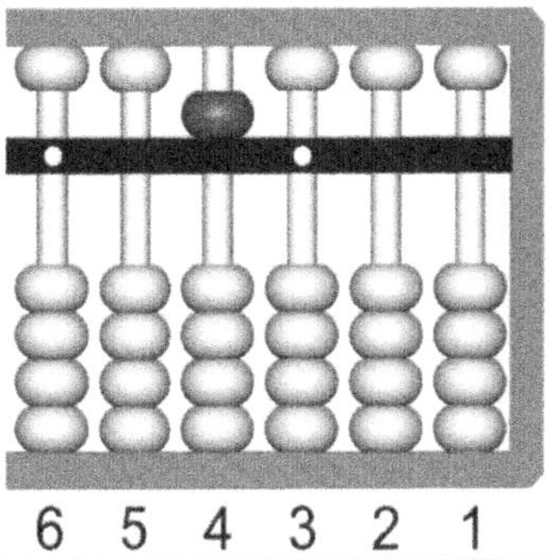

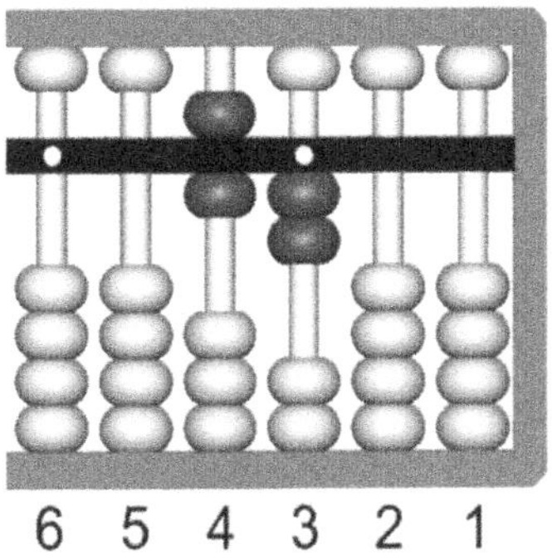

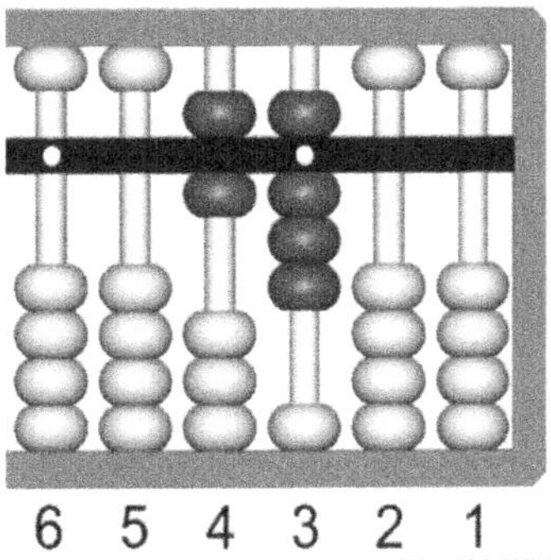

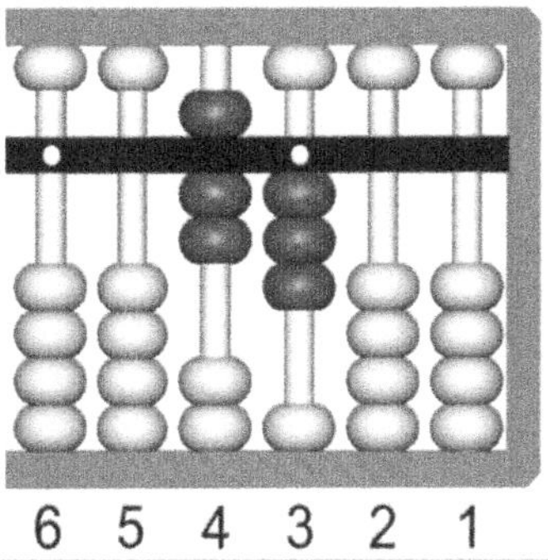

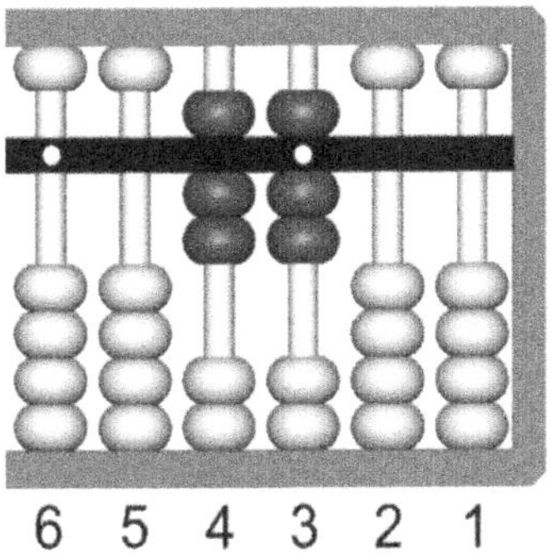

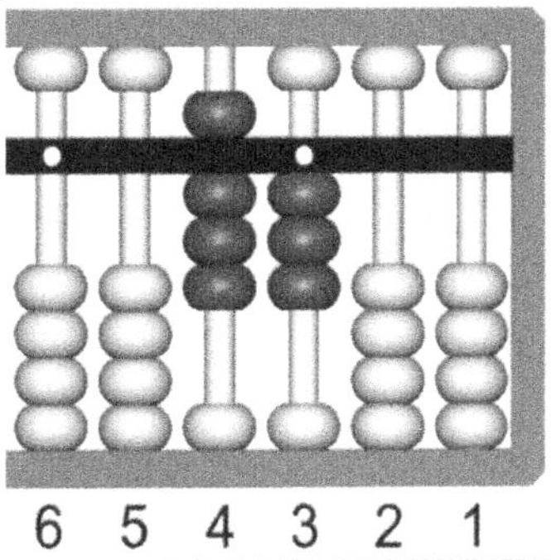

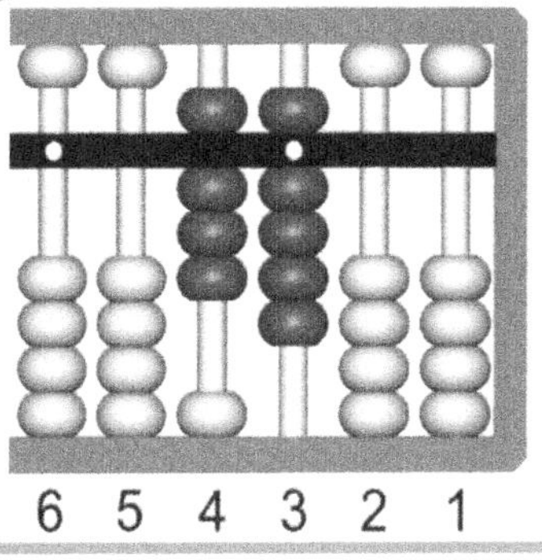

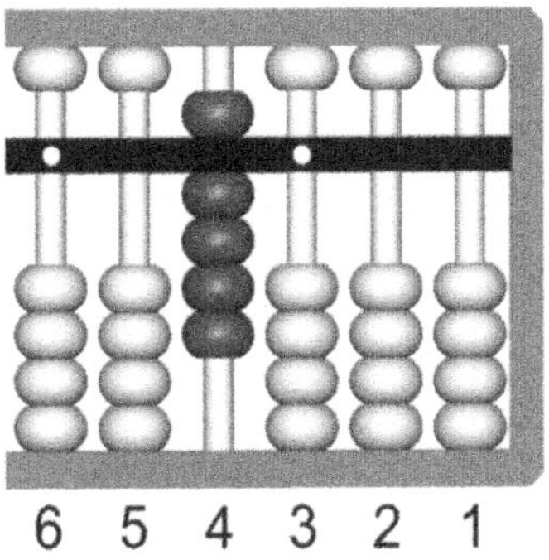

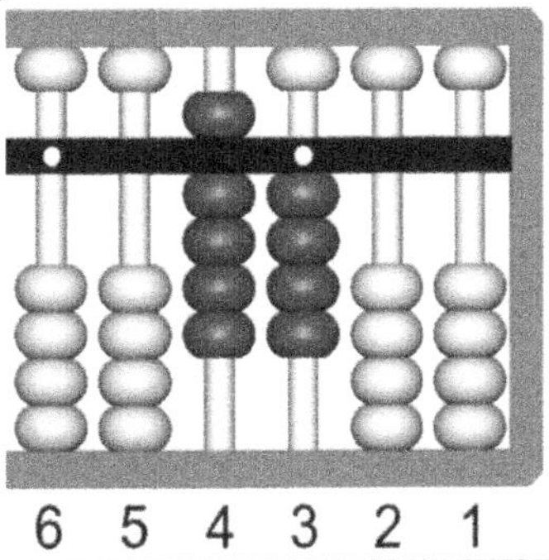

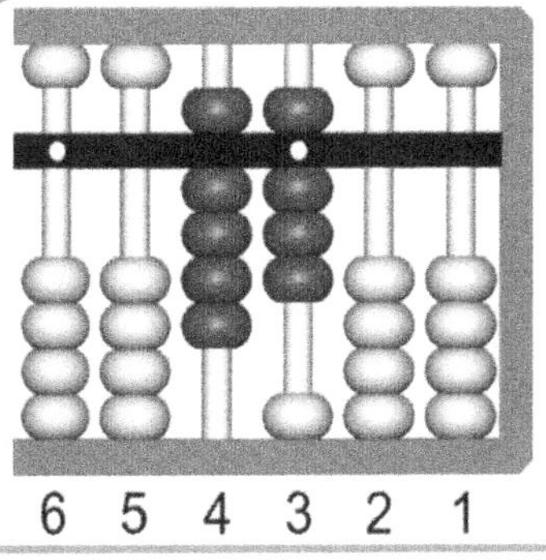

Time to use the **instruction book!** Go to instruction book **page 27.**

WORKBOOK WORK - 3

(Answers to workbook work 3 are on pages 78 to 81)

Add the numbers with your abacus and write the answer in the white box.

Examples:

Add these together

Write the answer here

	1			2			14
	4			4			21
=	5		=	6		=	35

---

①		②		③		④	
	1		1		1		2
	3		5		8		2
=		=		=		=	

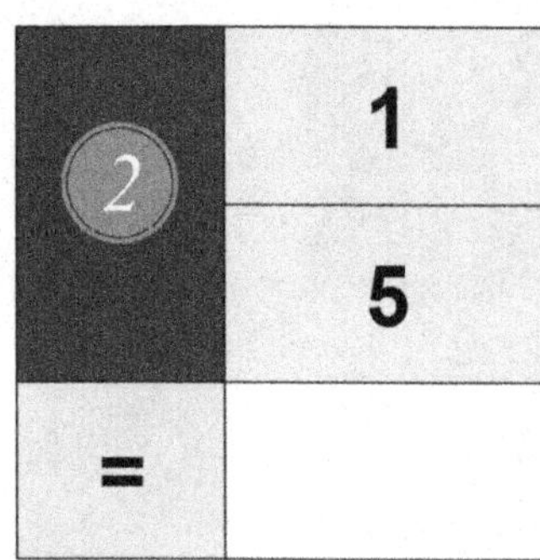
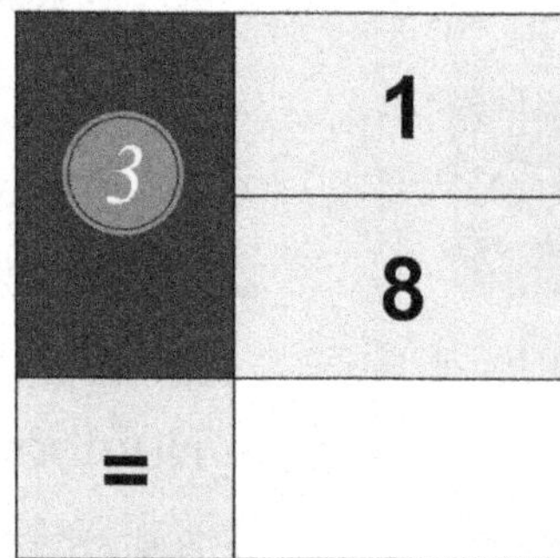

⑤		⑥		⑦		⑧	
	2		2		3		3
	3		6		3		6
=		=		=		=	

⑨		⑩		⑪		⑫	
	4		4		1		5
	5		4		8		2
=		=		=		=	

WORKBOOK - 3

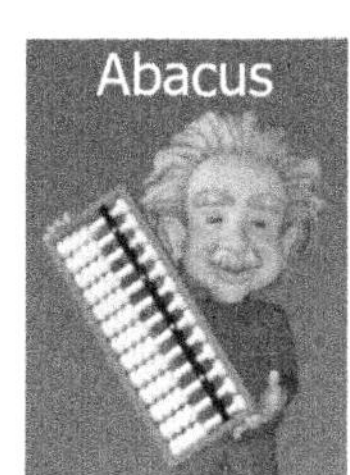

13	14	15	16
5	6	6	7
4	2	3	1
=	=	=	=

17	18	19	20
7	4	8	6
2	2	1	1
=	=	=	=

21	22 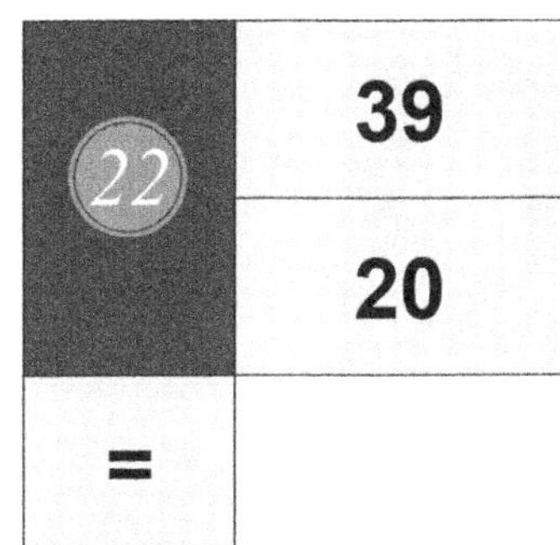	23	24
29	39	91	10
10	20	4	5
=	=	=	=

25	26	27	28
10	10	11	11
7	10	14	25
=	=	=	=

29	30	31	32
12	12	13	13
12	5	13	35
=	=	=	=

WORKBOOK - 3

No.			
33	18	11	=
34	14	14	=
35	24	12	=
36	26	3	=
37	35	21	=
38	30	14	=
39	40	12	=
40	52	4	=
41	57	12	=
42	67	11	=
43	60	9	=
44	62	14	=
45	74	25	=
46	75	3	=
47	70	9	=
48	80	13	=
49	86	11	=
50	82	4	=
51	91	4	=
52	94	5	=

WORKBOOK WORK - 3

2 Find the correct column for the digit, by putting a circle around the column number.

Examples:

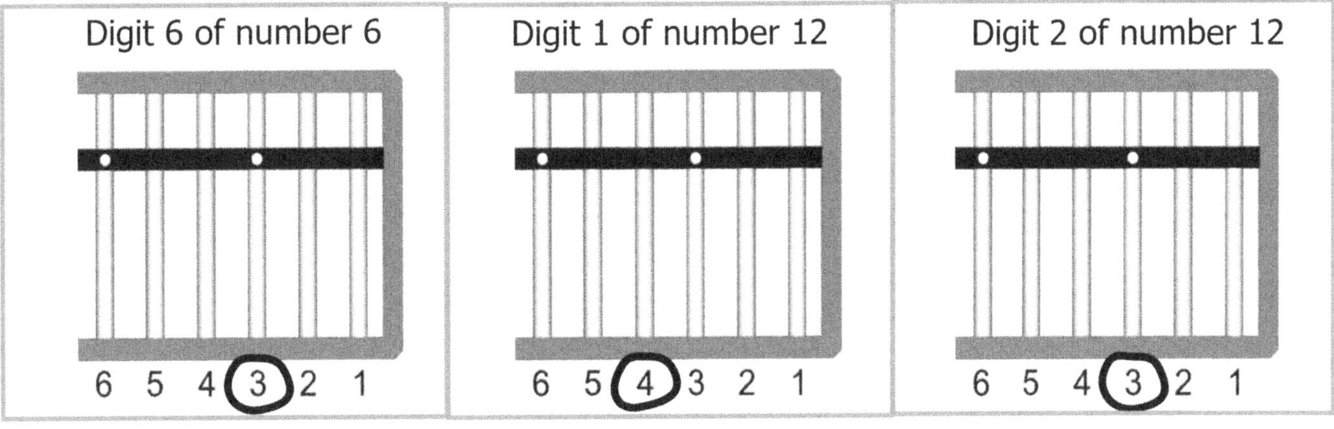

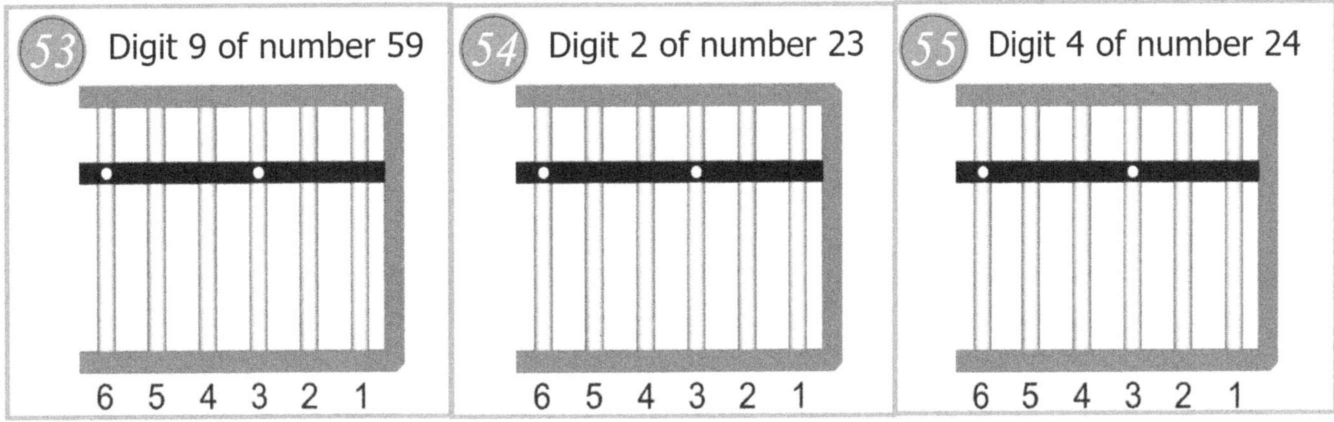

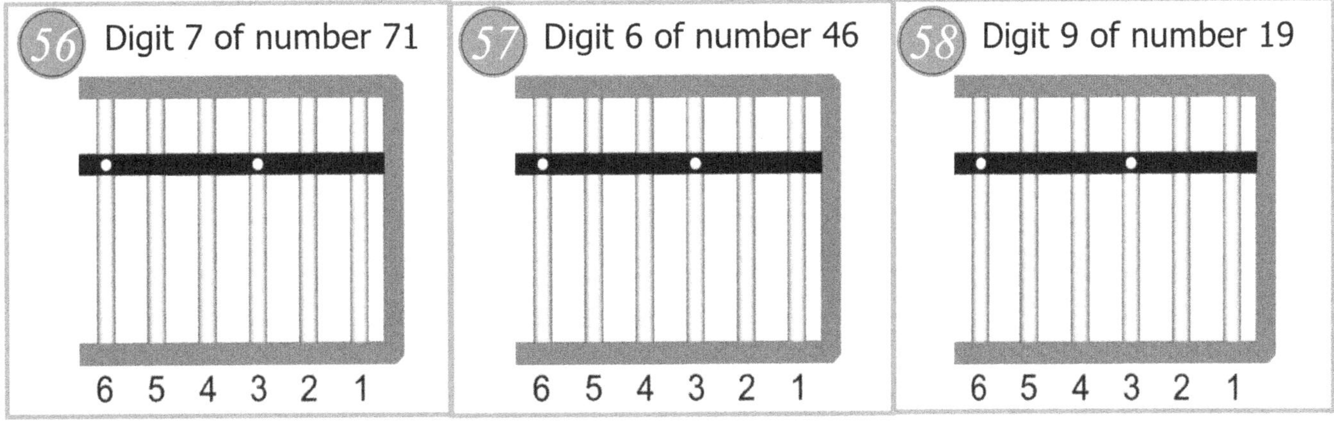

WORKBOOK WORK - 3

59 Digit 1 of number 10

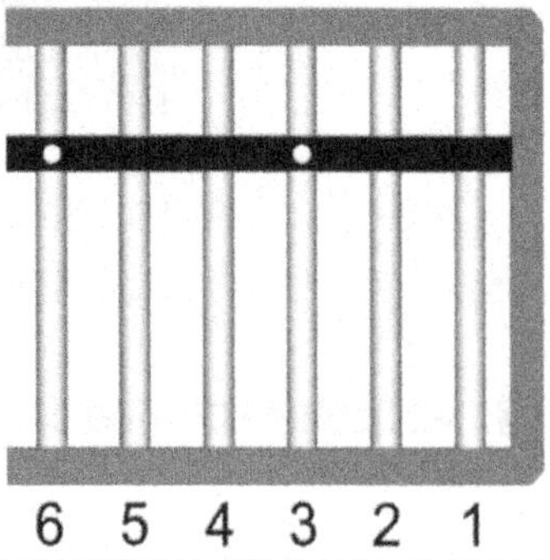

6 5 4 3 2 1

60 Digit 2 of number 12

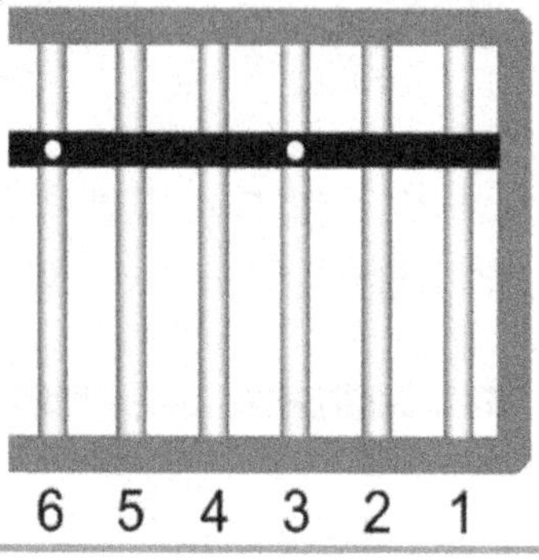

6 5 4 3 2 1

61 Digit 7 of number 74

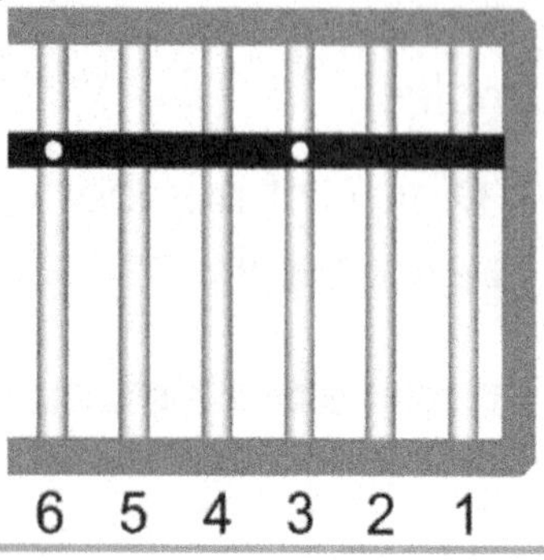

6 5 4 3 2 1

62 Digit 3 of number 13

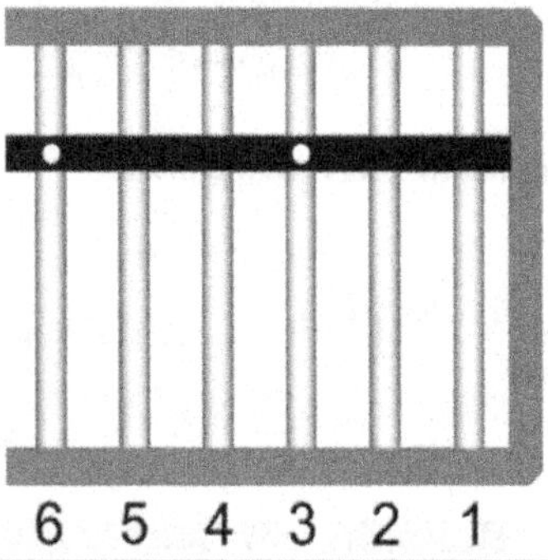

6 5 4 3 2 1

63 Digit 7 of number 17

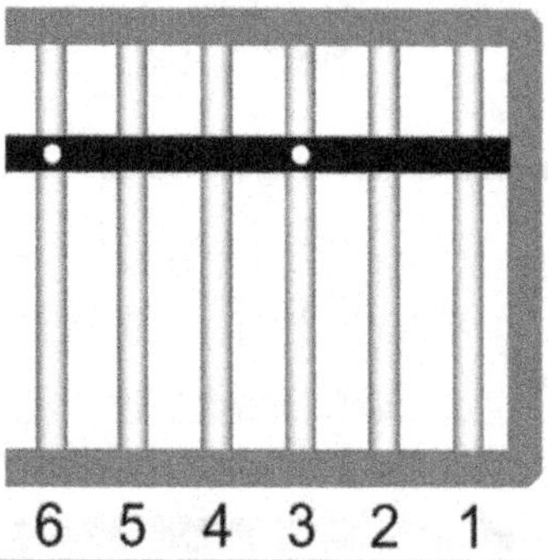

6 5 4 3 2 1

64 Digit 8 of number 80

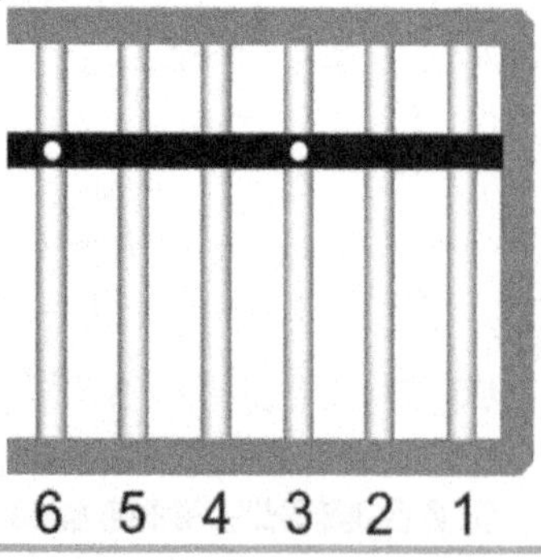

6 5 4 3 2 1

65 Digit 3 of number 3

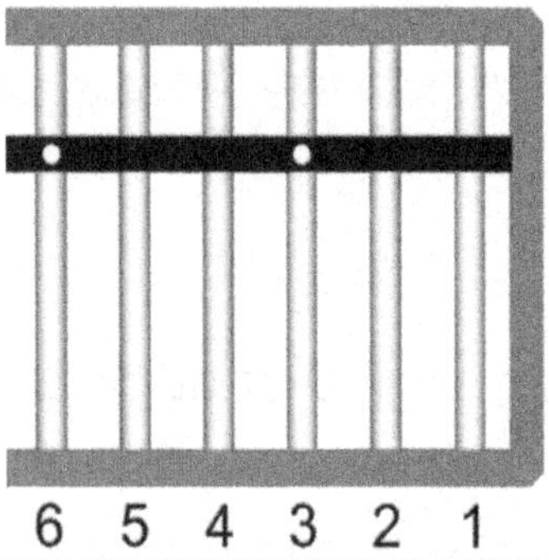

6 5 4 3 2 1

66 Digit 4 of number 45

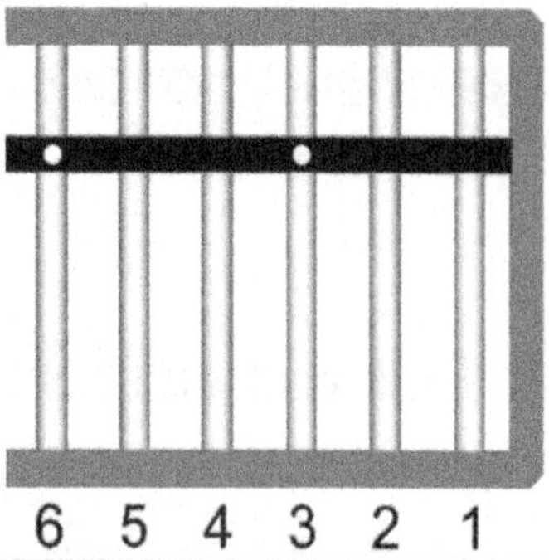

6 5 4 3 2 1

67 Digit 9 of number 91

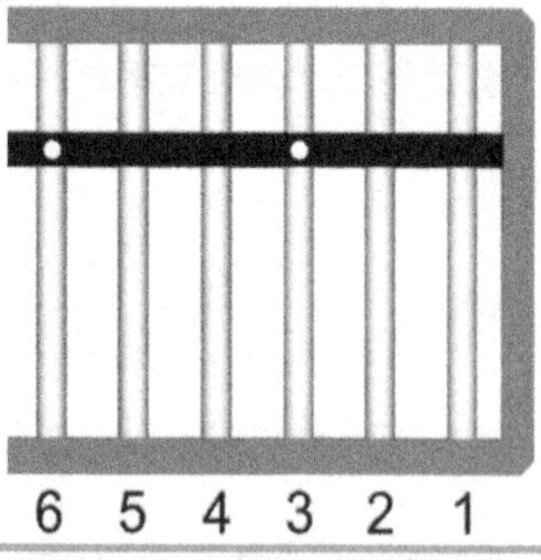

6 5 4 3 2 1

Time to use the **instruction book!** Go to instruction book **page 33.**

WORKBOOK WORK - 4

(Answers to workbook work 4 are on pages 82 to 87)

1 Add the numbers with your abacus and write the answer in the white box.

Examples: Add these together — Write the answer here

	1
	4
=	5

	2
	4
=	6

	14
	21
=	35

1	
	1
	1
=	

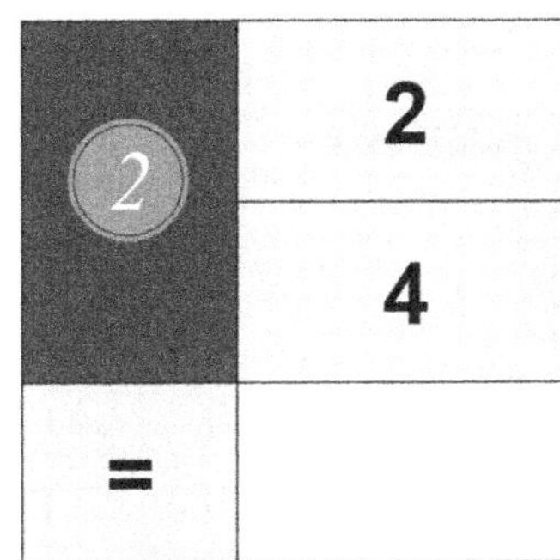

2	
	2
	4
=	

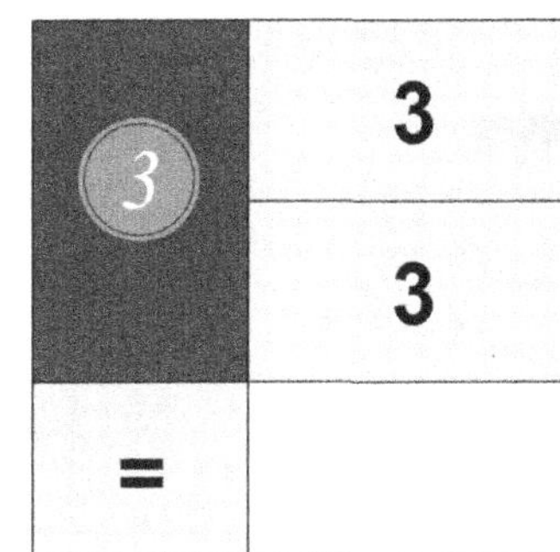

3	
	3
	3
=	

4	
	4
	4
=	

5	
	2
	6
=	

6	
	10
	2
=	

7	
	15
	5
=	

8	
	20
	10
=	

9	
	32
	3
=	

10	
	40
	20
=	

11	
	45
	5
=	

12	
	50
	25
=	

WORKBOOK WORK - 4

2 Write down the number that is shown on the imaginary abacus representation.

Examples:

The number is: 29	The number is: 44	The number is: 61

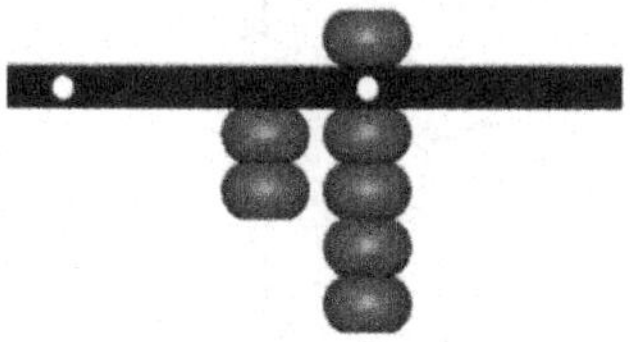

13 The number is:	14 The number is:	15 The number is:
16 The number is:	17 The number is:	18 The number is:
19 The number is:	20 The number is:	21 The number is:
22 The number is:	23 The number is:	24 The number is:

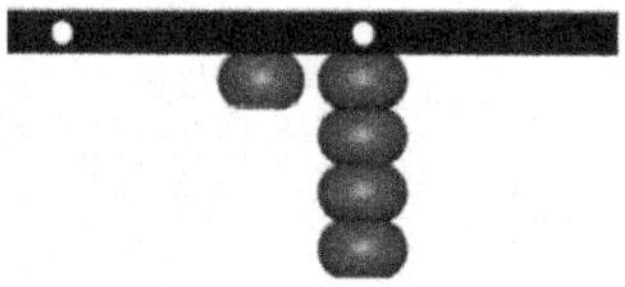

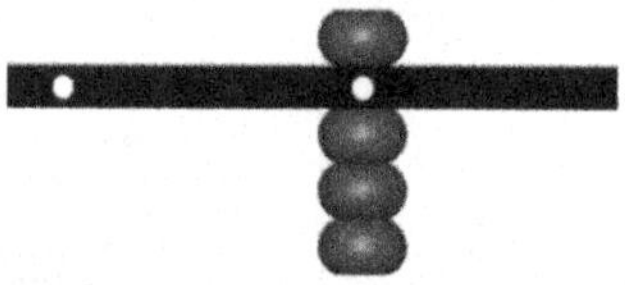

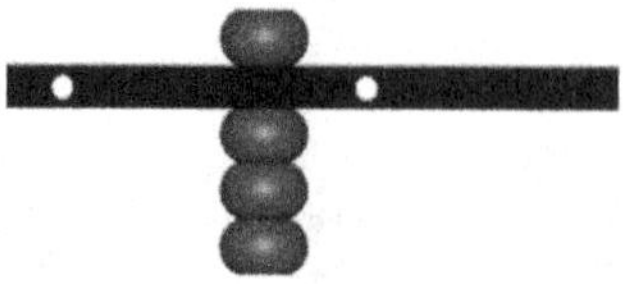

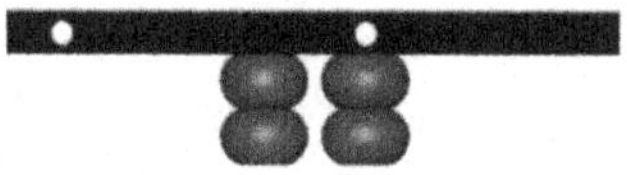

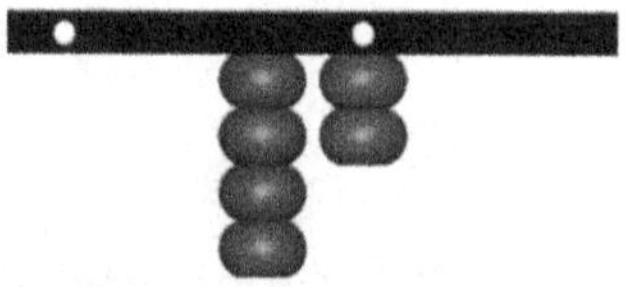

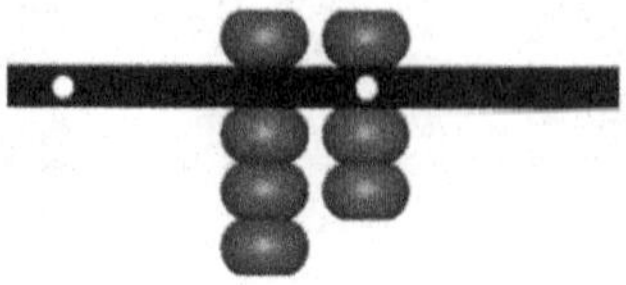

WORKBOOK WORK - 4

3 Draw the beads on the empty imaginary abacus representation to show the number given.

Examples:

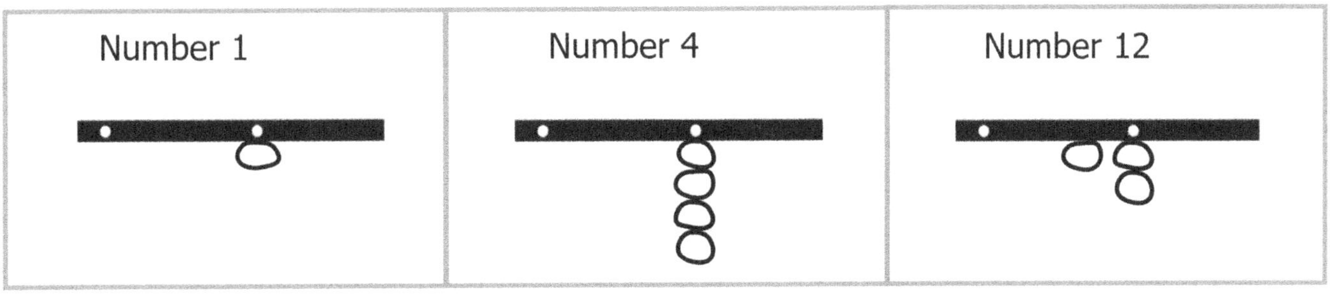

Number 1	Number 4	Number 12

25 Number 10	26 Number 25	27 Number 41
28 Number 9	29 Number 74	30 Number 12
31 Number 22	32 Number 78	33 Number 14
34 Number 95	35 Number 58	36 Number 66

WORKBOOK WORK - 4

37 Number 4

38 Number 55

39 Number 92

40 Number 2

41 Number 20

42 Number 33

43 Number 55

44 Number 63

45 Number 15

46 Number 7

47 Number 36

48 Number 39

49 Number 94

50 Number 57

51 Number 27

52 Number 82

53 Number 13

54 Number 88

Imagine

WORKBOOK WORK - 4

4 Add the numbers using an imaginary abacus and write the answer in the white box.

Examples:

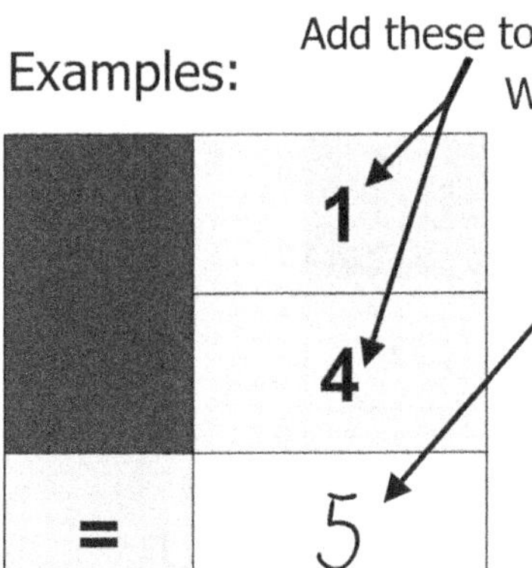

	1		2		14
	4		4		21
=	5	=	6	=	35

55	1	56	1	57	1	58	2
	3		5		8		2
=		=		=		=	

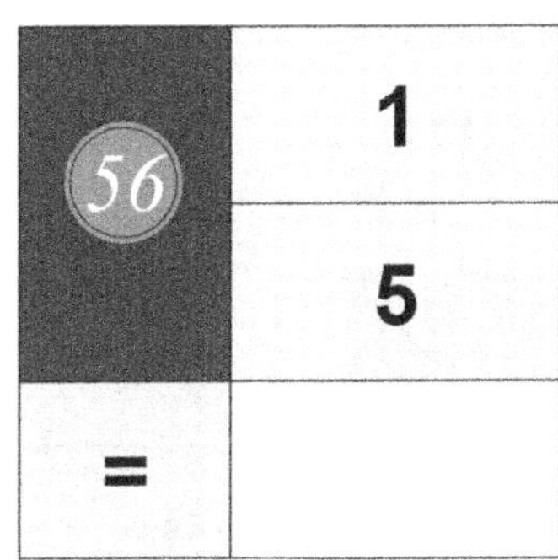

59	2	60	2	61	3	62	3
	3		9		3		6
=		=		=		=	

63	4	64	4	65	4	66	5
	5		7		9		5
=		=		=		=	

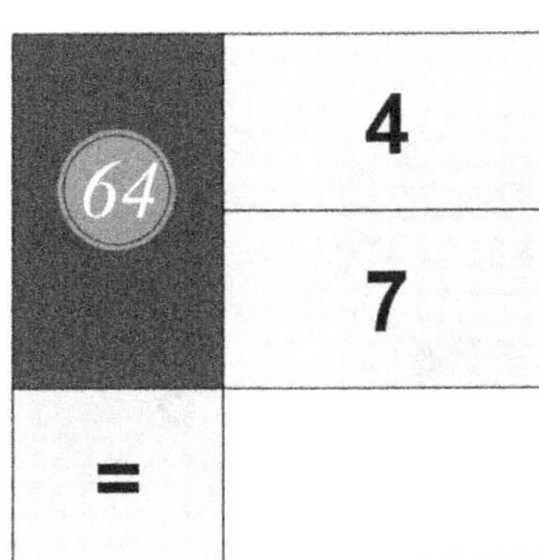

WORKBOOK - 4

No.			=
67	5	9	
68	6	5	
69	6	7	
70	7	1	
71	7	5	
72	8	4	
73	8	8	
74	8	9	
75	9	4	
76	9	6	
77	9	9	
78	10	5	
79	10	7	
80	10	10	
81	11	14	
82	11	25	
83	12	12	
84	12	5	
85	13	8	
86	13	35	

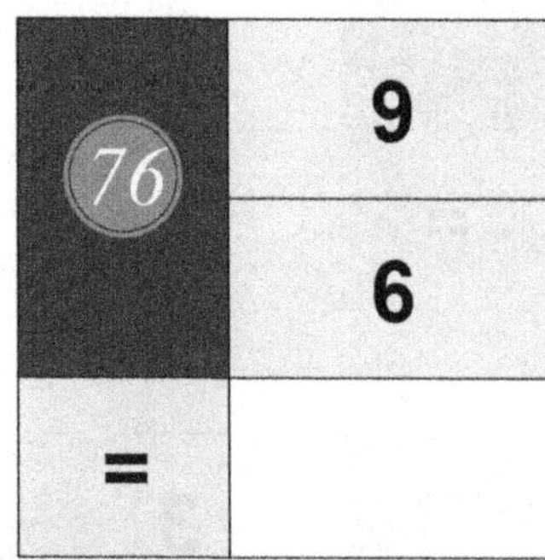

WORKBOOK - 4

No.			
87	18	9	=
88	19	14	=
89	24	12	=
90	26	7	=
91	35	25	=
92	38	14	=
93	40	12	=
94	52	9	=
95	57	9	=
96	67	14	=
97	68	9	=
98	69	14	=
99	74	25	=
100	75	3	=
101	78	9	=
102	80	13	=
103	86	11	=
104	89	4	=
105	91	4	=
106	94	5	=

Time to use the **instruction book!** Go to instruction book **page 40.**

WORKBOOK WORK - 5

(Answers to workbook work 5 are on pages 88 to 93)

 Add the numbers with your abacus and write the answer in the white box.

Examples:

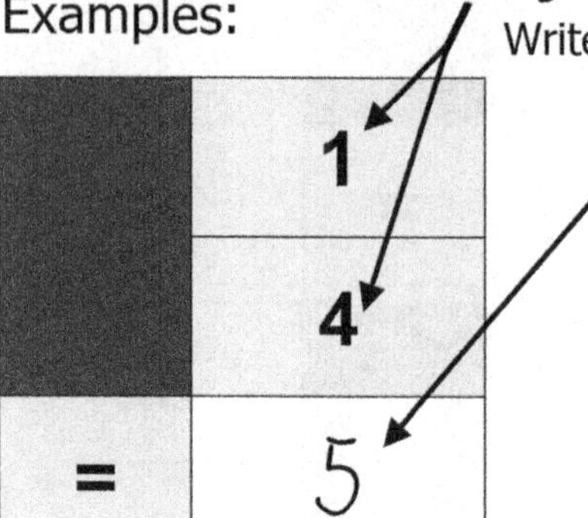

	1
	4
=	5

	2
	4
=	6

	14
	21
=	35

1	9
	2
=	

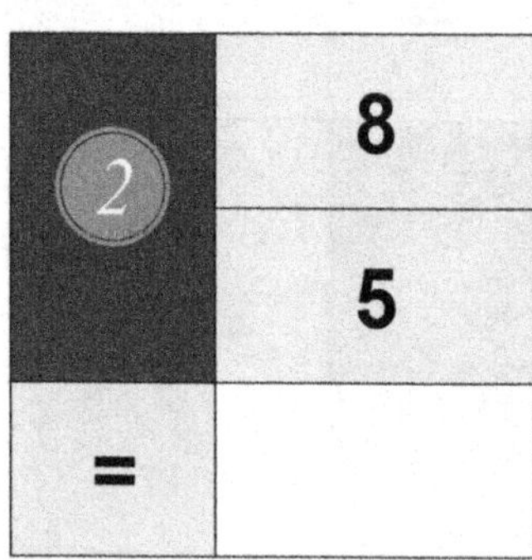

2	8
	5
=	

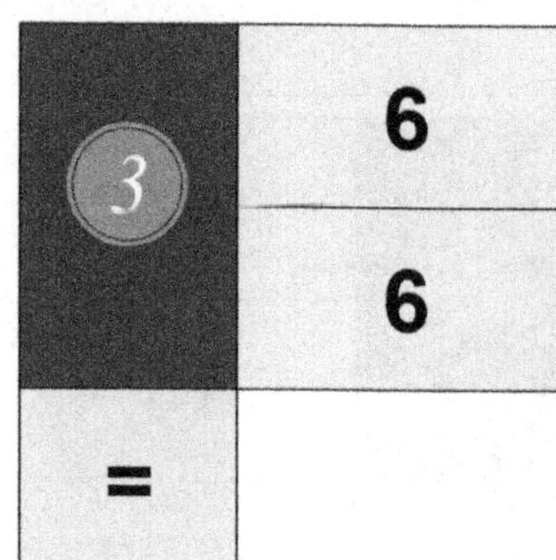

3	6
	6
=	

4	15
	6
=	

5	18
	3
=	

6	27
	9
=	

7	36
	8
=	

8	48
	6
=	

9	56
	5
=	

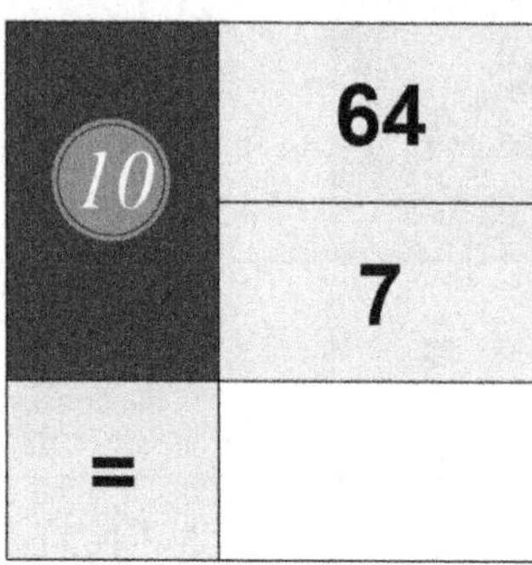

10	64
	7
=	

11	72
	9
=	

12	81
	9
=	

WORKBOOK - 5

No.		
13	57	9
=		

No.	First	Second	=
13	57	9	
14	11	9	
15	17	7	
16	78	12	
17	75	15	
18	8	4	
19	18	8	
20	63	9	
21	29	4	
22	39	6	
23	38	19	
24	19	5	
25	18	18	
26	34	16	
27	21	19	
28	36	25	
29	44	17	
30	9	9	
31	13	7	
32	54	36	

WORKBOOK - 5

No.	First	Second	=
33	100	55	
34	123	123	
35	110	105	
36	222	111	
37	324	322	
38	400	123	
39	555	105	
40	780	120	
41	111	98	
42	145	125	
43	132	122	
44	456	123	
45	910	18	
46	821	135	
47	666	258	
48	471	234	
49	744	147	
50	632	122	
51	482	135	
52	321	123	

WORKBOOK WORK - 5

2 Add the numbers using an imaginary abacus and write the answer in the white box.

Examples:

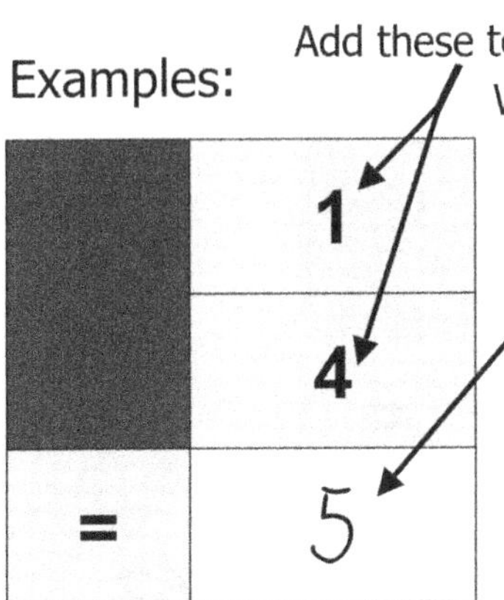

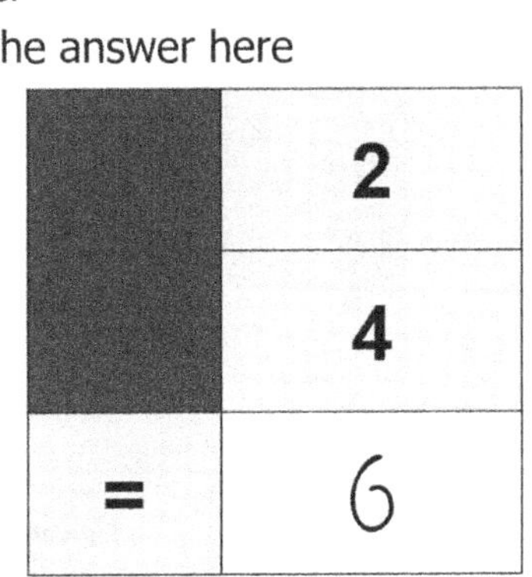

	14
	21
=	35

53	
	9
	2
=	

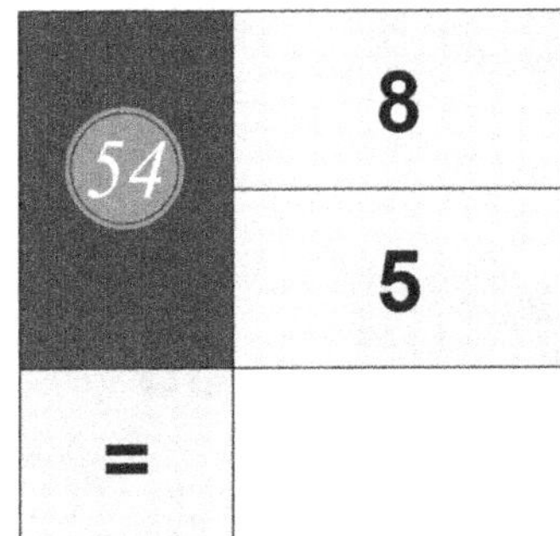

54	
	8
	5
=	

55	
	6
	6
=	

56	
	15
	6
=	

57	
	18
	3
=	

58	
	17
	9
=	

59	
	16
	8
=	

60	
	18
	6
=	

61	
	26
	5
=	

62	
	25
	7
=	

63	
	14
	9
=	

64	
	9
	9
=	

WORKBOOK - 5

No.	First number	Second number	=
65	18	9	
66	19	5	
67	27	7	
68	26	9	
69	35	8	
70	25	9	
71	22	9	
72	52	9	
73	57	7	
74	67	5	
75	68	9	
76	69	6	
77	77	4	
78	75	6	
79	78	9	
80	76	5	
81	58	3	
82	34	9	
83	14	9	
84	33	8	

WORKBOOK - 5

No.		
85	13	9
=		
86	15	5
=		
87	75	7
=		
88	82	9
=		
89	66	8
=		
90	35	9
=		
91	12	9
=		
92	53	9
=		
93	67	7
=		
94	75	5
=		
95	82	9
=		
96	93	6
=		
97	56	4
=		
98	66	6
=		
99	77	9
=		
100	88	5
=		
101	17	3
=		
102	22	9
=		
103	32	9
=		
104	43	8
=		

Time to use the **instruction book!** Go to instruction book **page 46.**

WORKBOOK WORK - 6

(Answers to workbook work 6 are on pages 94 to 98)

Abacus

1 Add the numbers with your abacus and write the answer in the white box.

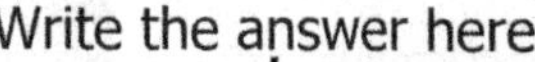

Add these together

Examples:

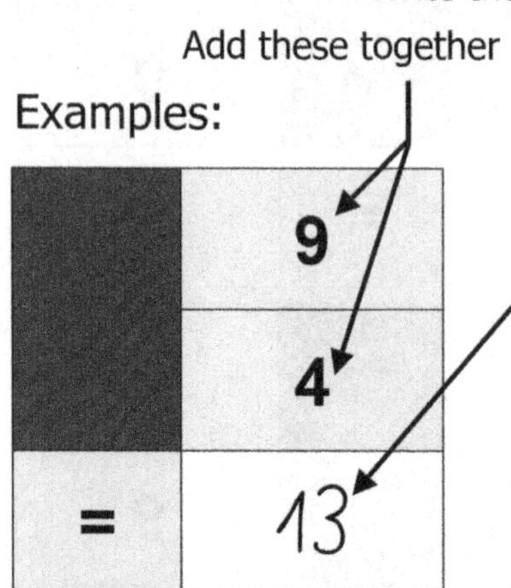

	9
	4
=	13

	122
	111
	33
=	266

	50
	35
	100
	55
=	240

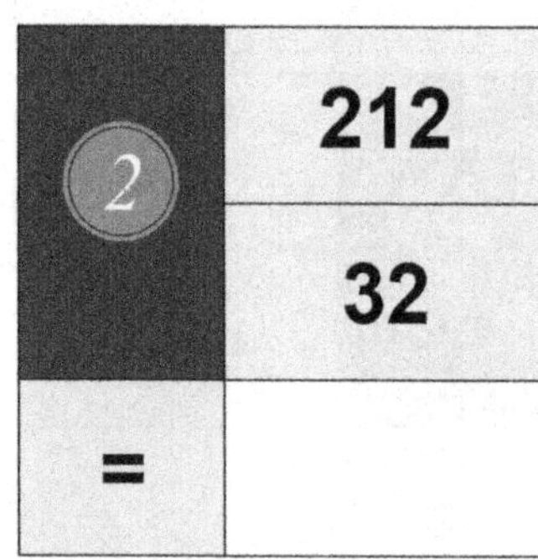

1	125
	20
=	

2	212
	32
=	

3	56
	56
=	

4	412
	12
=	

5	245
	145
=	

6	74
	58
=	

7	345
	123
=	

8	499
	111
=	

9	293
	195
=	

10	99
	1
=	

11	199
	99
=	

12	722
	133
=	

WORKBOOK - 6

No.		
13	512	36	=
14	823	119	=
15	699	111	=
16	442	36	=
17	85	85	=
18	99	9	=
19	180	19	=
20	325	32	=
21	15	6	=
22	19	9	=
23	29	29	=
24	91	90	=
25	99	19	=
26	77	16	=
27	652	321	=
28	336	25	=
29	55	55	=
30	442	38	=
31	256	123	=
32	321	321	=

WORKBOOK - 6

33	
	20
	5
	40
=	

34	
	35
	10
	20
=	

35	
	45
	55
	31
=	

36	
	90
	4
	16
=	

37	
	56
	14
	80
=	

38	
	5
	12
	36
=	

39	
	8
	18
	34
=	

40	
	100
	55
	15
=	

41	
	250
	150
	100
	125
=	

42	
	50
	21
	40
	19
=	

43	
	23
	17
	69
	11
=	

44	
	156
	4
	25
	15
	103
=	

WORKBOOK - 6

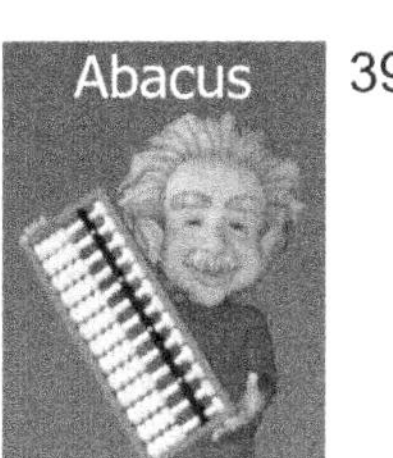

45	
	25
	8
	17
=	

46	
	365
	32
	11
=	

47	
	74
	6
	150
=	

48	
	14
	6
	9
=	

49	
	102
	36
	8
	14
=	

50	
	250
	12
	32
	40
=	

51	
	10
	32
	6
	85
=	

52	
	6
	14
	154
	730
=	

53	
	7
	11
	40
	23
	99
=	

54	
	101
	124
	123
	210
	310
=	

55	
	50
	25
	13
	19
	99
=	

56	
	378
	21
	9
	12
	3
=	

WORKBOOK WORK - 6

2 Add the numbers using an imaginary abacus and write the answer in the white box.

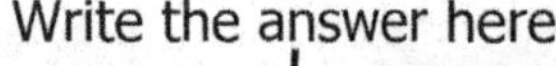

Add these together

Examples:

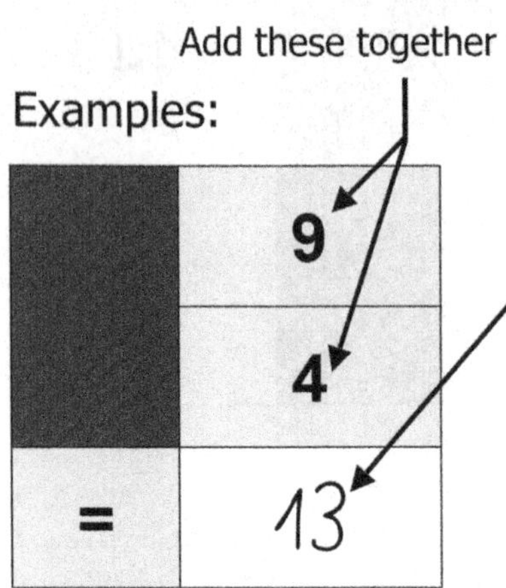

	122
	111
	33
=	266

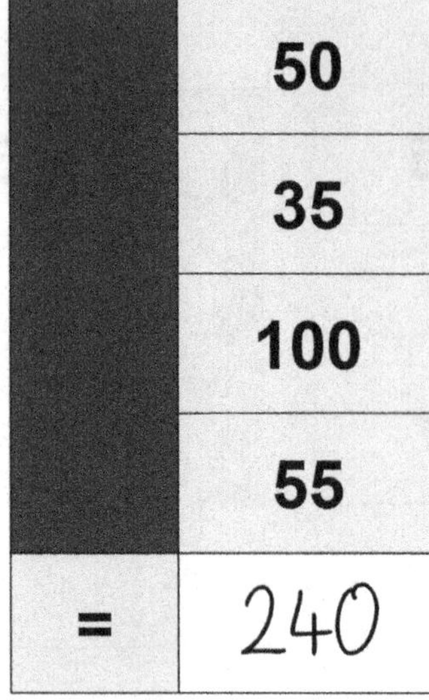

57	99
	1
=	

58	29
	14
=	

59	23
	11
=	

60	35
	5
=	

61	65
	9
=	

62	99
	10
=	

63	123
	32
=	

64	100
	55
=	

65	222
	111
=	

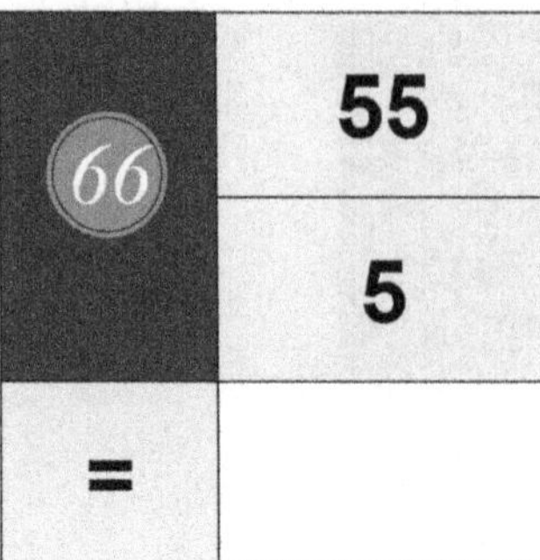

66	55
	5
=	

67	444
	222
=	

68	750
	122
=	

WORKBOOK - 6

No.			
69	40	25	=
70	78	12	=
71	65	15	=
72	105	25	=
73	74	26	=
74	49	14	=
75	80	36	=
76	110	35	=
77	120	110	=
78	82	18	=
79	63	17	=
80	75	15	=
81	96	4	=
82	150	44	=
83	34	8	=
84	450	150	=
85	250	50	=
86	48	22	=
87	35	30	=
88	750	225	=

WORKBOOK - 6

89	
	150
	25
	10
=	

90	
	253
	17
	30
=	

91	
	115
	3
	24
=	

92	
	111
	89
	10
=	

93	
	4
	156
	230
=	

94	
	62
	40
	18
=	

95	
	32
	32
	32
=	

96	
	85
	2
	110
=	

97	
	12
	10
	18
	3
=	

98	
	30
	50
	5
	15
=	

99	
	60
	12
	10
	7
=	

100	
	20
	15
	65
	5
	100
=	

Time to use the **instruction book!** Go to instruction book **page 51.**

WORKBOOK WORK - 7

(Answers to workbook work 7 are on pages 99 to 102)

Draw the beads on the empty abacus to represent the number given.

Examples:

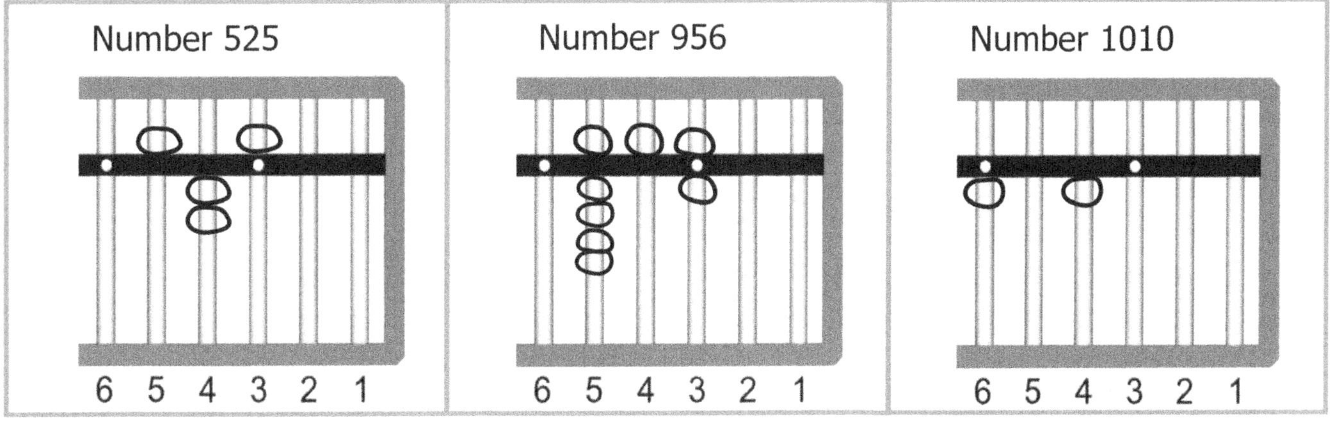

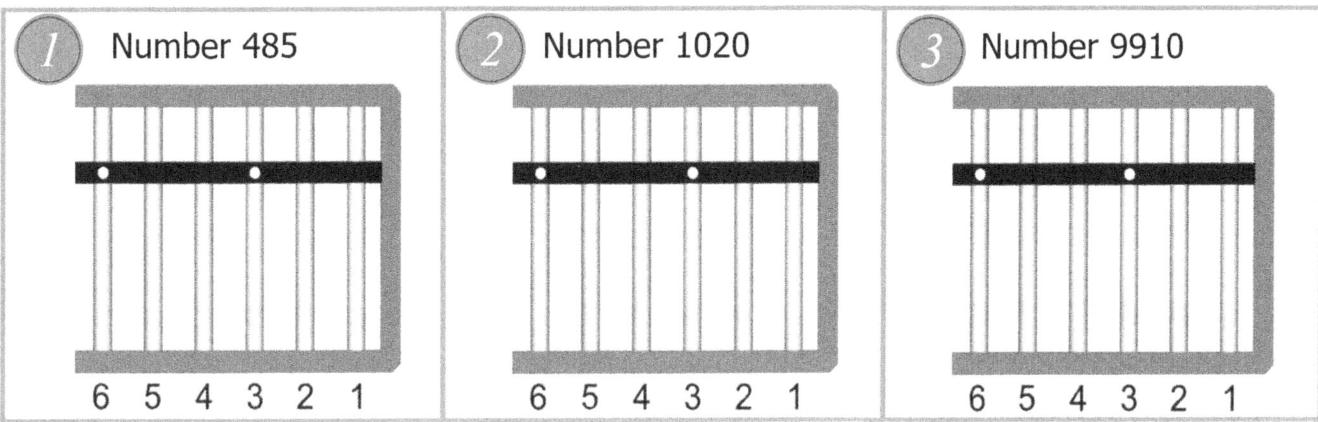

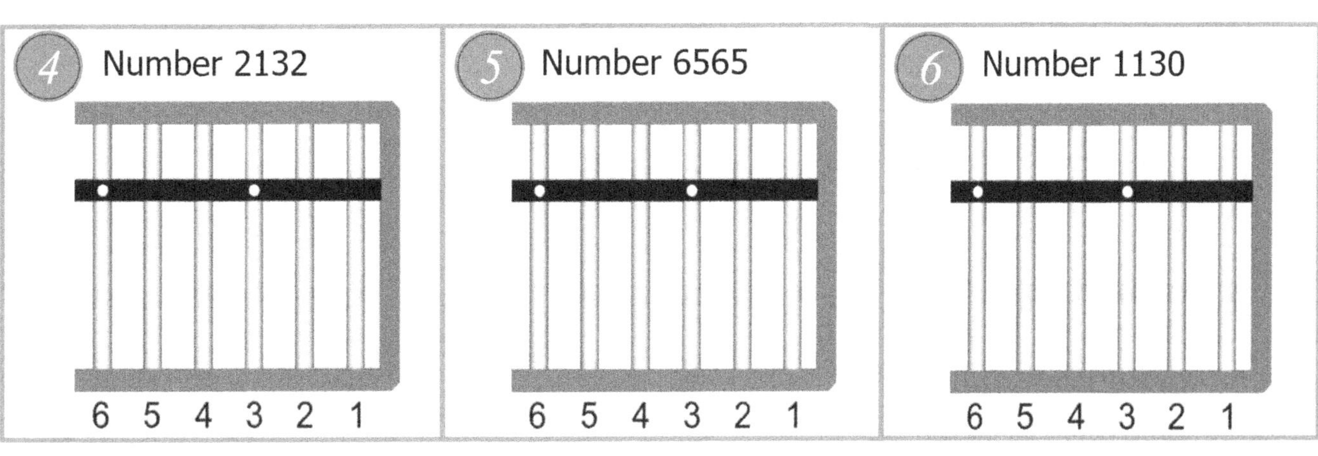

WORKBOOK WORK - 7

7 Number 523

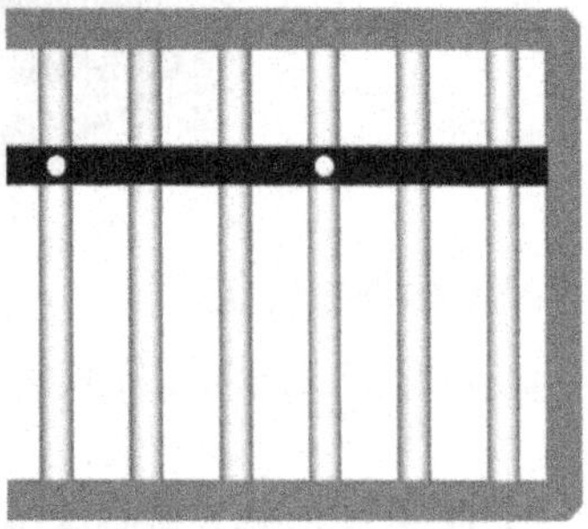

6 5 4 3 2 1

8 Number 801

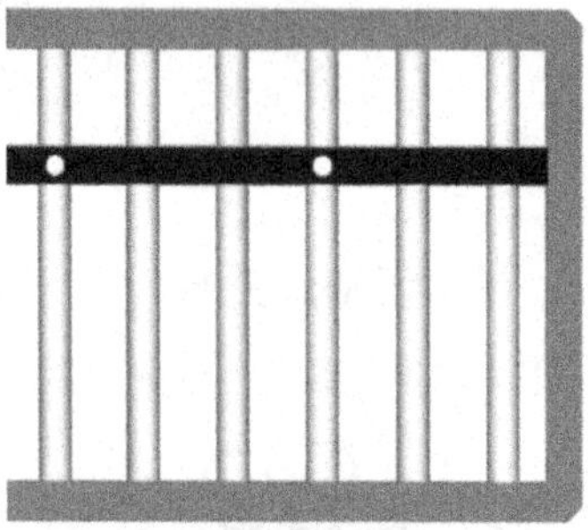

6 5 4 3 2 1

9 Number 333

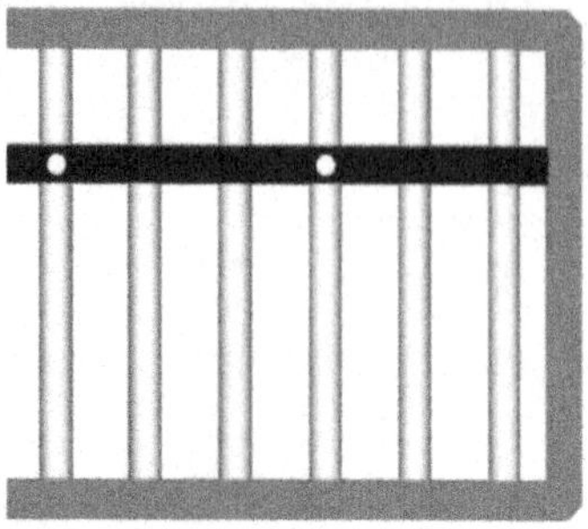

6 5 4 3 2 1

10 Number 1111

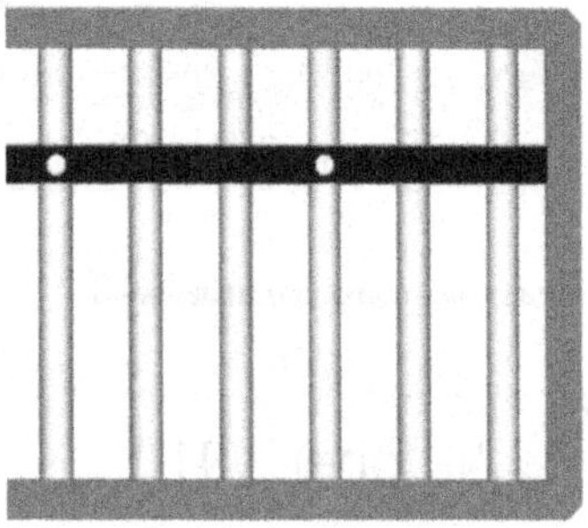

6 5 4 3 2 1

11 Number 5555

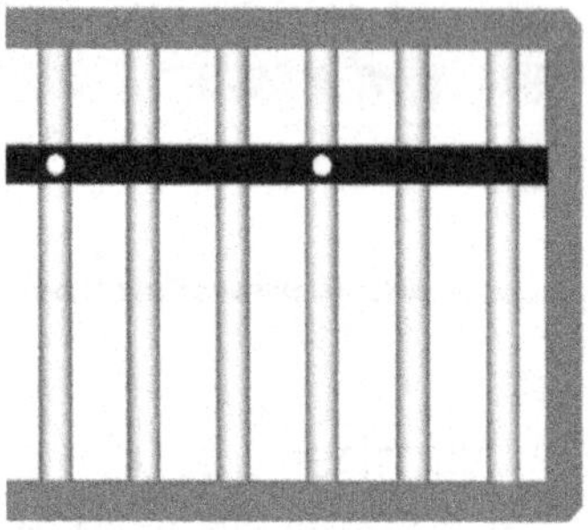

6 5 4 3 2 1

12 Number 400

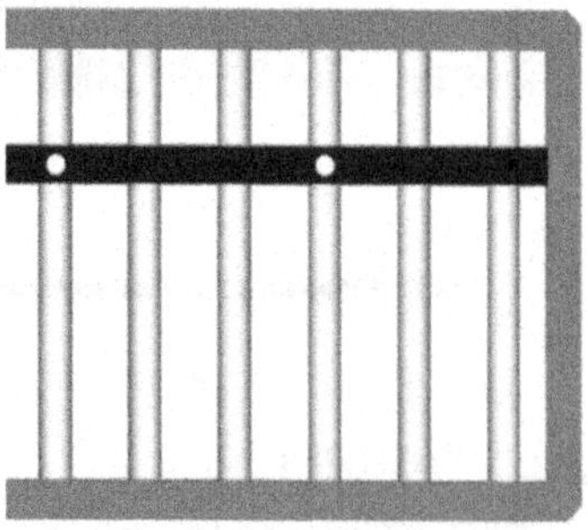

6 5 4 3 2 1

13 Number 4000

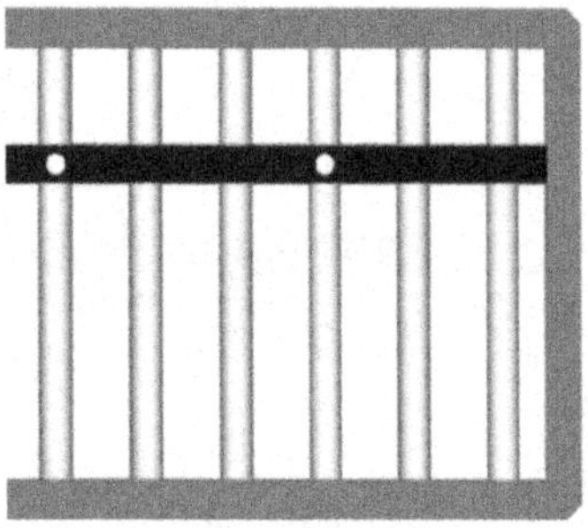

6 5 4 3 2 1

14 Number 40

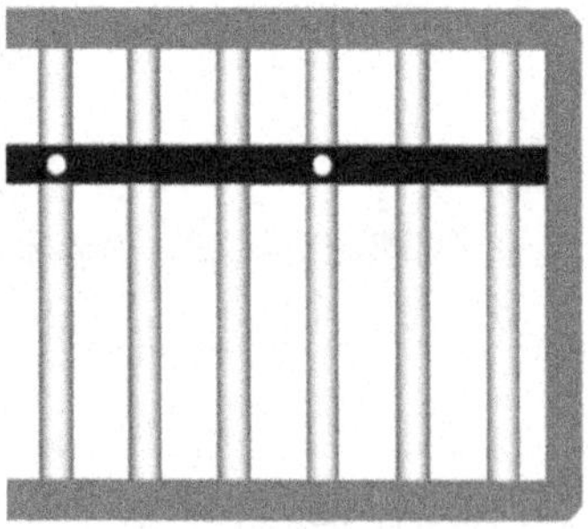

6 5 4 3 2 1

15 Number 2121

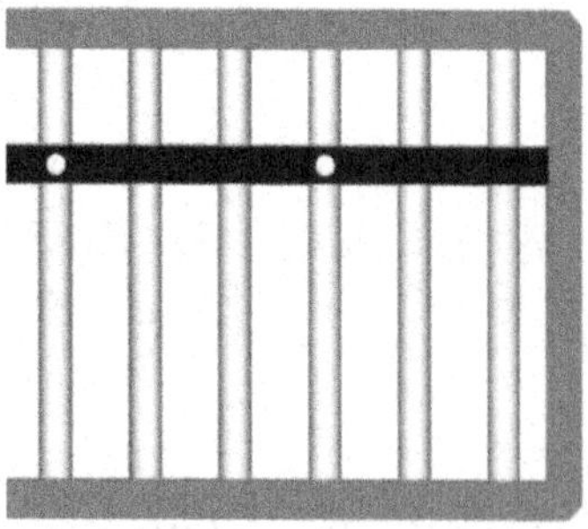

6 5 4 3 2 1

16 Number 3216

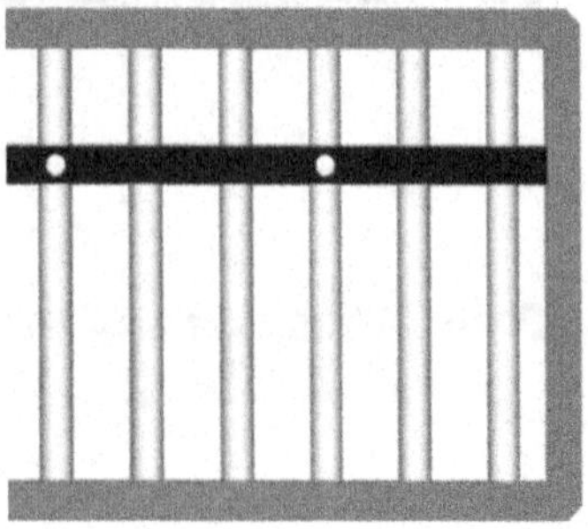

6 5 4 3 2 1

17 Number 9000

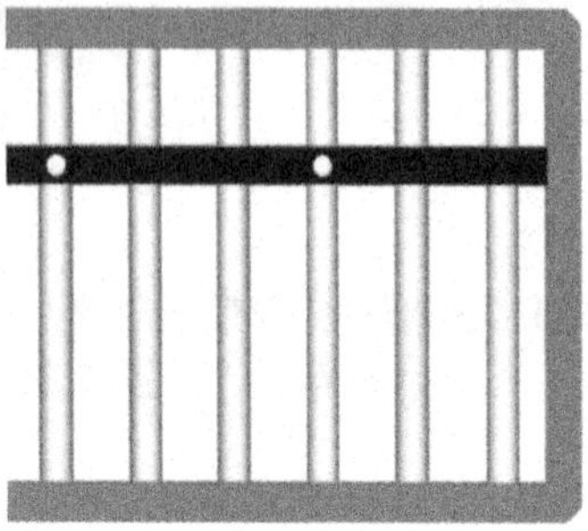

6 5 4 3 2 1

18 Number 1234

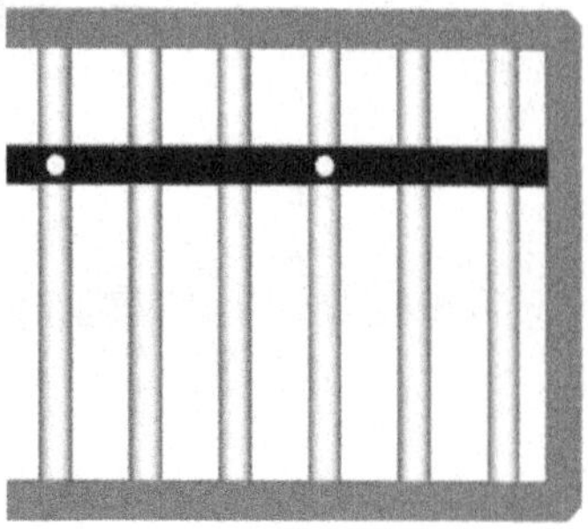

6 5 4 3 2 1

WORKBOOK WORK - 7

2 Find the correct column for the digit, by putting a circle around the column number.

Examples:

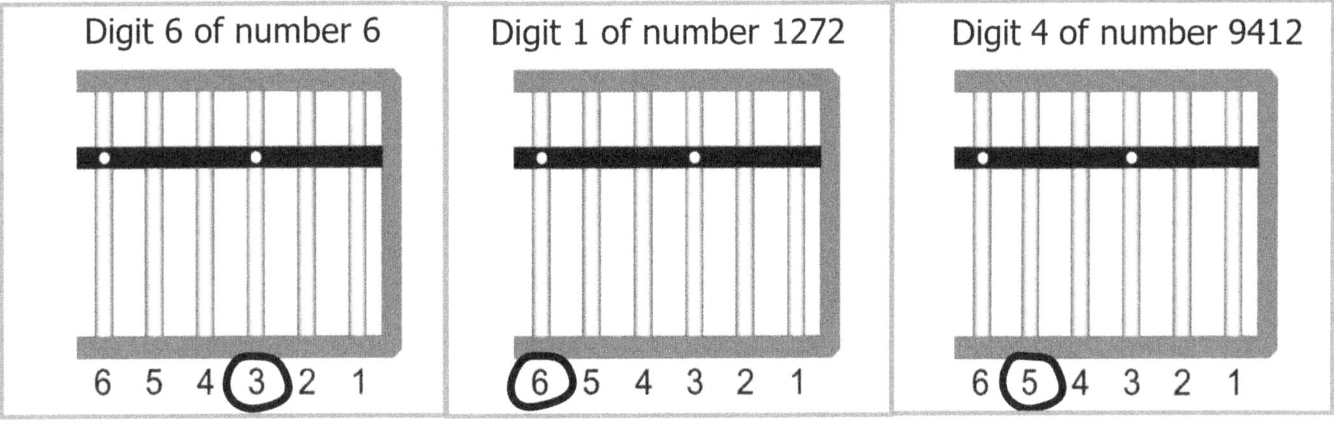

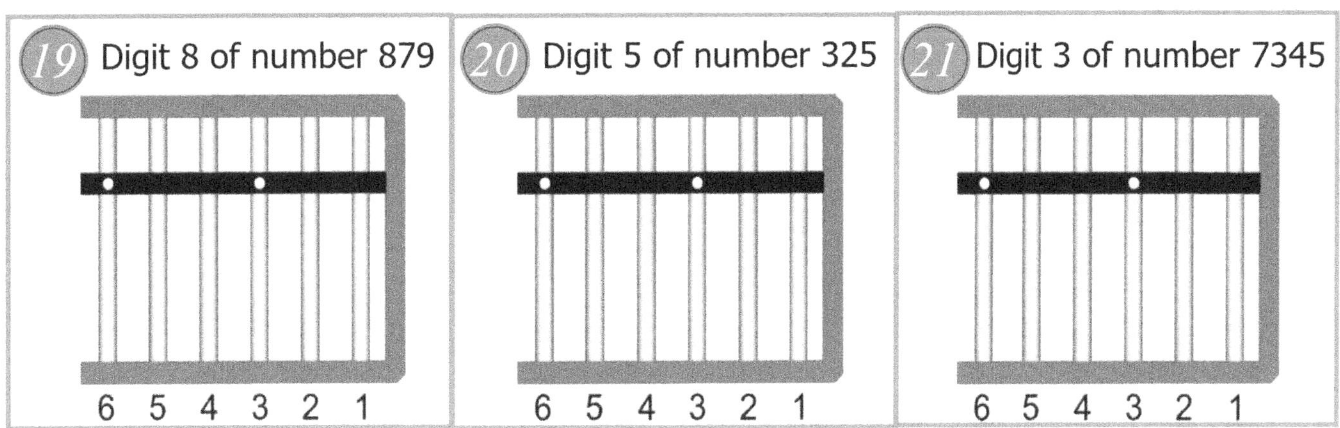

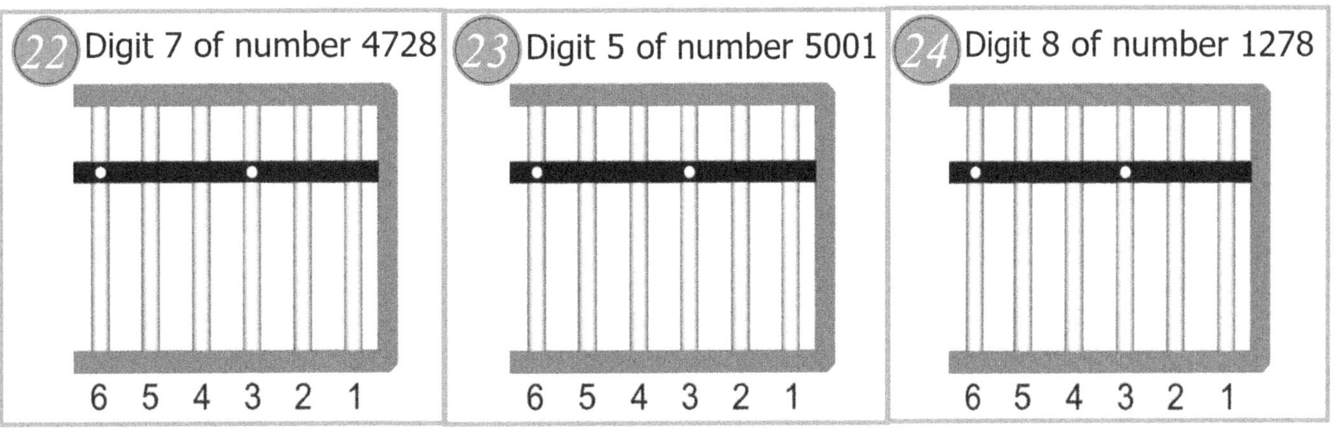

WORKBOOK WORK - 7

3 Write down the number that is shown on the abacus.

Examples:

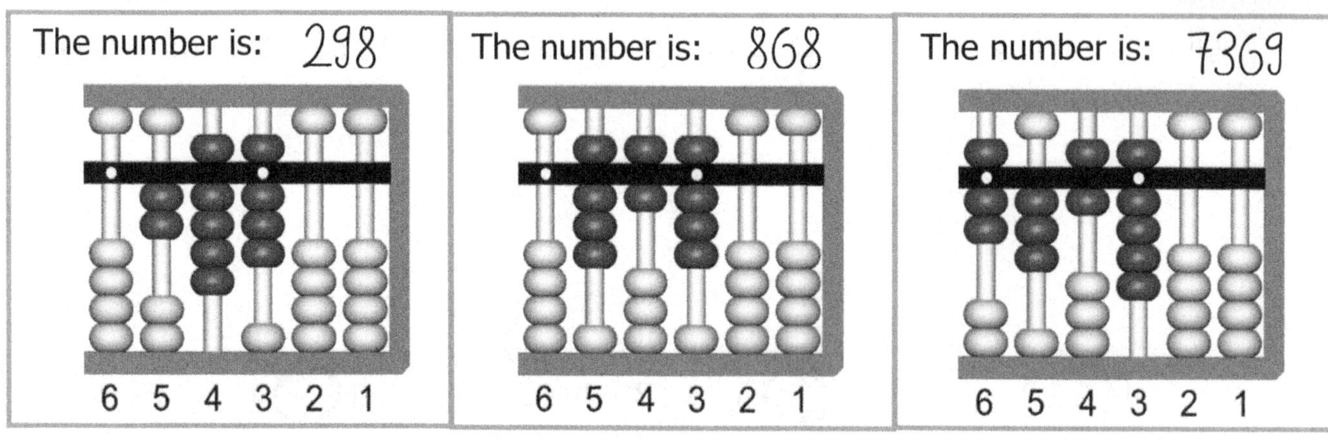

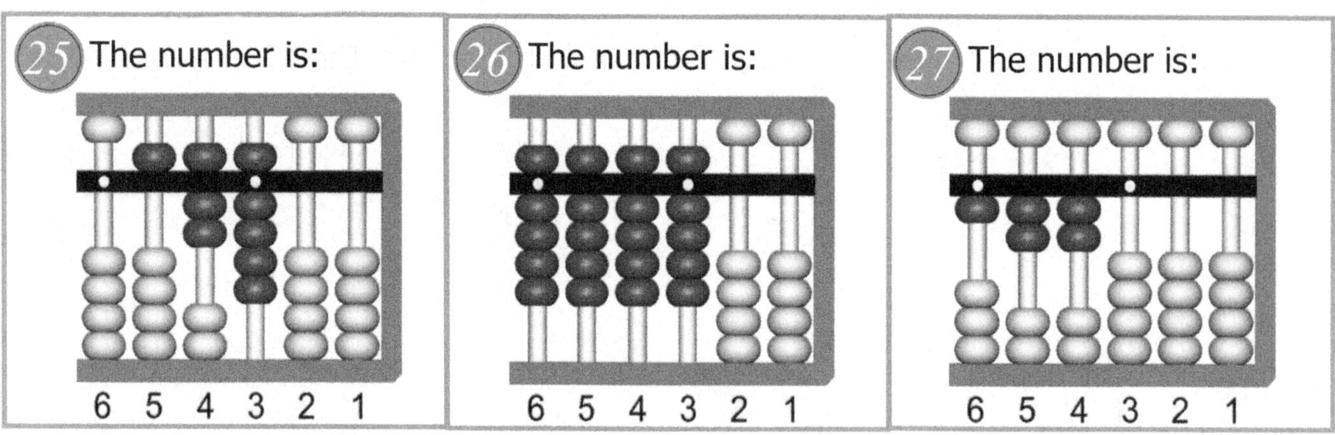

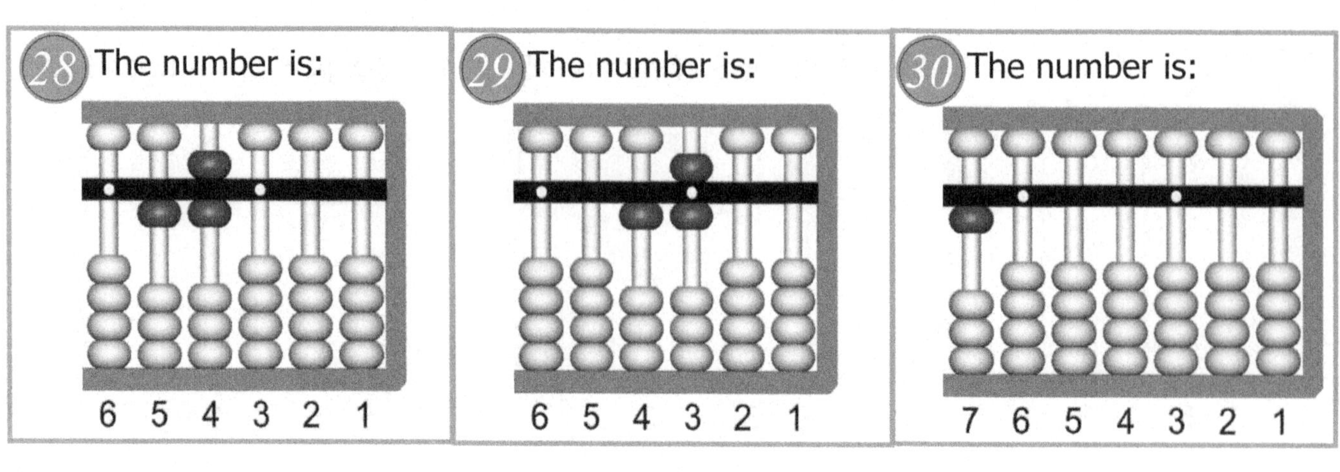

WORKBOOK WORK - 7

4 Subtract the numbers with your abacus and write the answer in the white box.

Examples:

Subtract these numbers

Write the answer here

	20		102		2354
	-4		-9		-235
=	16	=	93	=	2119

31	26	32	58	33	123	34	454
	-3		-12		-23		-10
=		=		=		=	

35	797	36	815	37	6655	38	365
	-43		–5		-122		-62
=		=		=		=	

39	995	40	155	41	6777	42	488
	-605		-23		-244		-244
=		=		=		=	

WORKBOOK WORK - 7

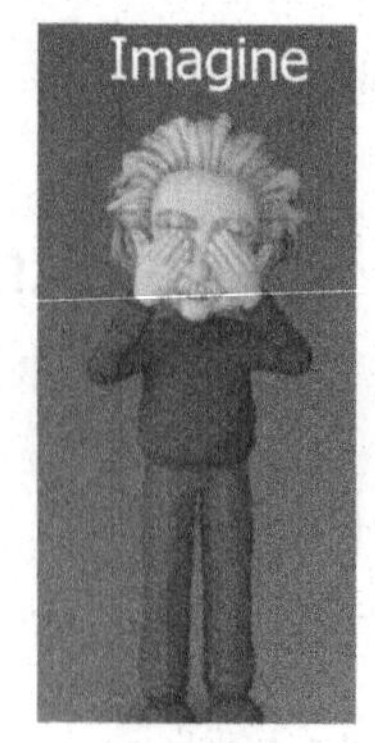

5 Subtract the numbers using an imaginary abacus and write the answer in the white box.

Examples:

Subtract these numbers

Write the answer here

	5		12		144
	-3		-4		-22
=	2	=	8	=	122

43	9	44	55	45	125	46	66
	-5		-5		-13		-33
=		=		=		=	

47	89	48	478	49	633	50	456
	-34		-423		-502		-323
=		=		=		=	

51	888	52	25	53	196	54	99
	-464		-12		-150		-66
=		=		=		=	

WORKBOOK - 7

55: 46 -25 =	56: 78 -55 =	57: 65 -15 =	58: 185 -25 =
59: 445 -201 =	60: 688 -354 =	61: 596 -73 =	62: 110 -10 =
63: 120 -110 =	64: 88 -18 =	65: 267 -13 =	66: 775 -70 =
67: 396 -202 =	68: 656 -133 =	69: 999 -131 =	70: 450 -150 =
71: 256 -152 =	72: 158 -136 =	73: 321 -111 =	74: 750 -620 =

Time to use the **instruction book!** Go to instruction book **page 58.**

WORKBOOK WORK - 8

(Answers to workbook work 8 are on pages 103 to 104)

1 Subtract the numbers with your abacus and write the answer in the white box.

Examples:

Subtract these numbers

Write the answer here

	20
	-4
=	16

	112
	-9
=	103

	2354
	-235
=	2119

1	26
	-7
=	

2	58
	-19
=	

3	123
	-24
=	

4	454
	-64
=	

5	771
	-192
=	

6	887
	–19
=	

7	6455
	-18
=	

8	333
	-17
=	

9	201
	-16
=	

10	133
	-15
=	

11	6033
	-100
=	

12	174
	-19
=	

WORKBOOK WORK - 8

2 Subtract the numbers using an imaginary abacus and write the answer in the white box.

Examples:

Subtract these numbers

Write the answer here

	5
	-3
=	2

	12
	-4
=	8

	144
	-22
=	122

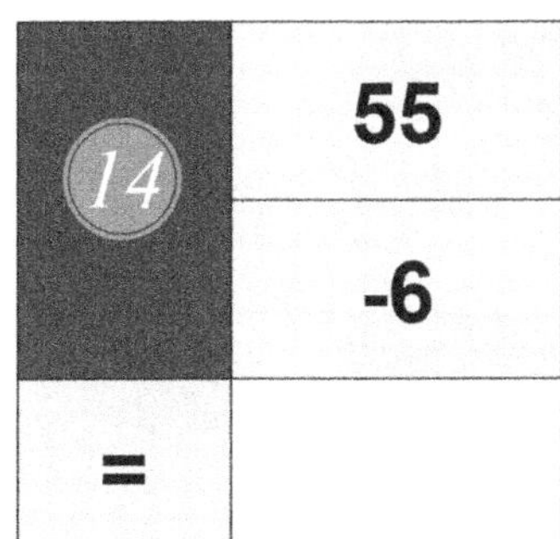

13	14	15	16
13	55	125	66
-4	-6	-33	-7
=	=	=	=

17	18	19	20
82	48	63	45
-13	-9	-54	-36
=	=	=	=

21	22	23	24
18	25	12	92
-9	-16	-8	-8
=	=	=	=

WORKBOOK - 8

25	46	
	-7	
=		

26	78	
	-9	
=		

27	65	
	-6	
=		

28	185	
	-86	
=		

29	45	
	-6	
=		

30	68	
	-9	
=		

31	56	
	-7	
=		

32	113	
	-9	
=		

33	120	
	-30	
=		

34	88	
	-9	
=		

35	20	
	-1	
=		

36	25	
	-7	
=		

37	96	
	-87	
=		

38	100	
	-20	
=		

39	45	
	-16	
=		

40	450	
	-360	
=		

Time to use the **instruction book!** Go to instruction book **page 65.**

WORKBOOK WORK - 9

(Answers to workbook work 9 are on pages 105 to 106)

1. Subtract the numbers with your abacus and write the answer in the white box.

Examples:

Subtract these numbers

Write the answer here

	Example	Example	Example
	20	112	2354
	-4	-9	-235
=	16	103	2119

	1	2	3	4
	163	100	772	1003
	-13	-5	-18	-8
=				

	5	6	7	8
	413	493	6546	606
	-56	-25	-109	-27
=				

	9	10	11	12
	117	1000	1010	100000
	-8	-13	-15	-1
=				

WORKBOOK WORK - 9

2 Subtract the numbers using an imaginary abacus and write the answer in the white box.

Examples:

Subtract these numbers

Write the answer here

	Example 1	Example 2	Example 3
	5	12	144
	-3	-4	-22
=	2	8	122

	13	14	15	16
	10	100	145	75
	-4	-6	-55	-7
=				

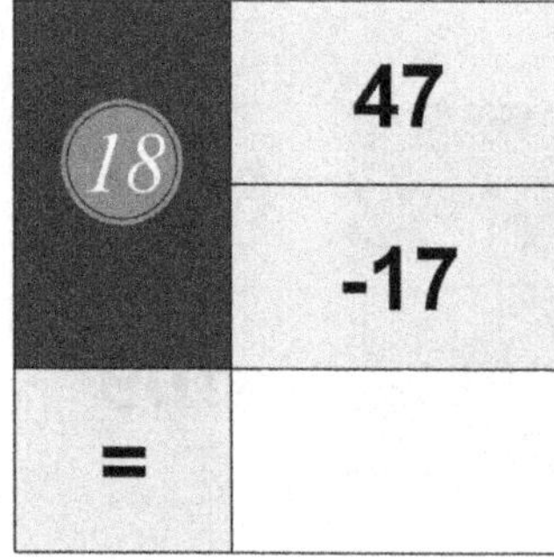

	17	18	19	20
	96	47	100	1000
	-6	-17	-5	-5
=				

	21	22	23	24
	444	10000	144	80
	-240	-1	-24	-8
=				

WORKBOOK - 9

25	55
	-15
=	

26	84
	-12
=	

27	63
	-11
=	

28	145
	-36
=	

29	125
	-6
=	

30	100
	-9
=	

31	200
	-9
=	

32	400
	-10
=	

33	555
	-7
=	

34	88
	-25
=	

35	165
	-125
=	

36	652
	-318
=	

37	241
	-56
=	

38	100
	-22
=	

39	599
	-222
=	

40	453
	-240
=	

Time to use the **instruction book!** Go to instruction book **page 68.**

WORKBOOK WORK - 10

(Answers to workbook work 10 are on pages 107 to 108)

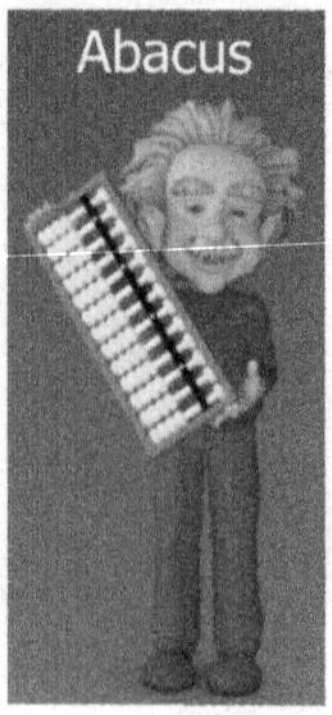

1 Calculate the result using your abacus and write the answer in the white box.

Find the result

Write the answer here, in the white box

Examples:

	25	15	-6	0	34	
	36	-9	12	-18	5	26

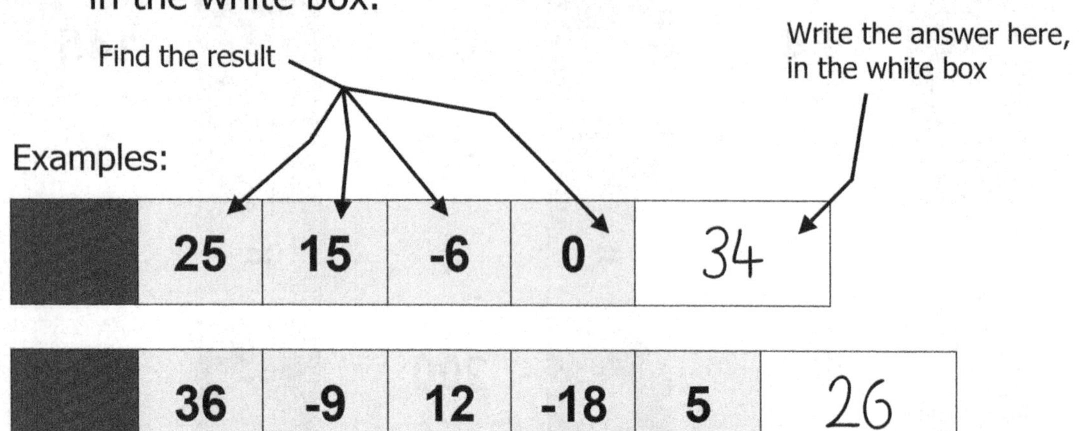

1	75	10	-15	-23			
2	85	-15	63	-35			
3	110	-35	20	-45			
4	254	125	45	-178	-52		
5	361	-117	-102	563			
6	855	-255	299	-320			
7	20	10	30	120	-80	104	
8	666	-333	44	-257			
9	810	189	-611				

WORKBOOK WORK - 10

2 Calculate the result using an imaginary abacus and write the answer in the white box.

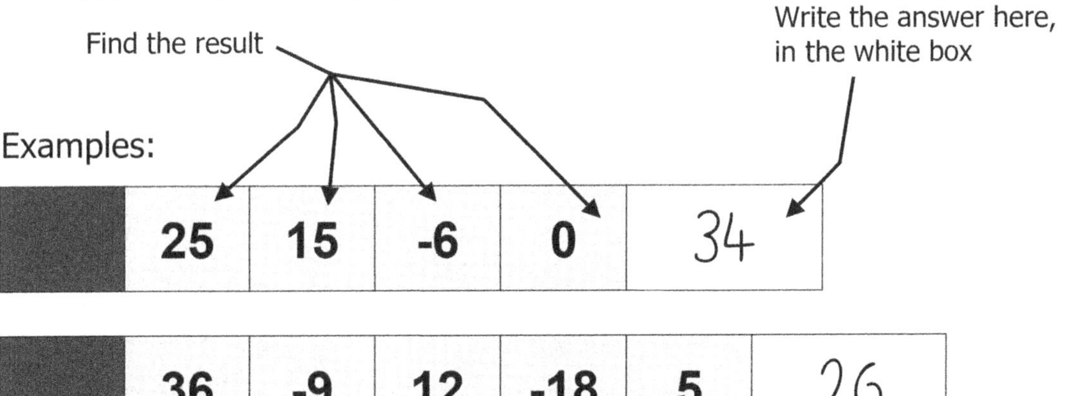

Examples:

	25	15	-6	0	34

	36	-9	12	-18	5	26

10	50	10	-15	-20	
11	85	-15	60	-35	
12	100	-15	20	-45	

13	254	100	-20	43	13	

14	42	-12	52	-22	
15	899	-202	-307	-300	

16	20	10	40	100	-65	-30	

Time to use the **instruction book!** Go to instruction book **page 75.**

WORKBOOK WORK - Reusable - Page 1

(Answers to reusable workbook work are on pages 109 to 122)

Abacus

A		
1	25	20
2	25	22
3	52	33
4	16	12
5	25	22
6	44	20
7	52	41
8	39	25
9	12	9
10	79	75
11	95	82
12	32	14
13	65	42
14	12	32
15	72	45
16	85	65
17	25	22
18	10	8
19	11	5
20	51	21
21	64	64
22	48	45
23	85	65
24	48	18
25	12	8
26	8	3
27	21	1
28	99	98
29	47	12
30	76	3

B		
1	65	55
2	23	22
3	60	32
4	26	12
5	14	12
6	44	41
7	65	52
8	63	37
9	74	15
10	85	75
11	88	82
12	36	18
13	54	42
14	26	32
15	85	20
16	55	25
17	92	6
18	75	12
19	76	11
20	65	55
21	66	64
22	49	22
23	85	66
24	91	18
25	21	10
26	26	5
27	26	6
28	52	23
29	58	47
30	56	26

C		
1	56	55
2	66	25
3	41	32
4	45	16
5	12	10
6	65	44
7	13	12
8	55	36
9	96	12
10	85	79
11	30	19
12	30	32
13	25	24
14	30	12
15	65	72
16	54	32
17	47	6
18	60	12
19	13	11
20	56	55
21	96	64
22	55	45
23	44	23
24	22	18
25	45	10
26	85	3
27	58	9
28	99	12
29	90	47
30	88	3

D		
1	82	55
2	28	25
3	42	32
4	32	22
5	82	22
6	25	18
7	76	42
8	33	32
9	21	20
10	55	25
11	30	19
12	36	32
13	65	24
14	12	12
15	72	72
16	85	32
17	25	6
18	39	36
19	55	54
20	51	26
21	64	23
22	32	30
23	92	5
24	75	6
25	76	20
26	85	65
27	58	8
28	99	14
29	90	47
30	77	11

WORKBOOK WORK - Reusable - Page 2

A		
1	21	55
2	31	25
3	41	32
4	18	22
5	25	36
6	45	18
7	52	66
8	32	32
9	12	12
10	77	25
11	95	21
12	31	32
13	65	25
14	12	13
15	72	77
16	85	32
17	85	6
18	10	66
19	11	54
20	52	26
21	64	23
22	48	32
23	85	9
24	78	6
25	12	22
26	8	55
27	21	8
28	99	15
29	88	47
30	76	12

B		
1	52	55
2	63	22
3	41	32
4	42	12
5	96	12
6	85	41
7	30	55
8	30	37
9	66	15
10	30	75
11	65	88
12	54	18
13	55	42
14	60	32
15	13	22
16	56	25
17	99	6
18	60	12
19	13	12
20	56	56
21	96	77
22	55	44
23	44	66
24	33	18
25	45	10
26	85	8
27	58	6
28	99	25
29	25	47
30	26	26

C		
1	56	65
2	66	23
3	41	60
4	45	26
5	12	14
6	65	44
7	13	65
8	55	63
9	96	74
10	85	85
11	30	88
12	30	36
13	25	54
14	30	26
15	65	85
16	54	55
17	47	92
18	60	75
19	13	76
20	56	65
21	96	66
22	55	49
23	44	85
24	22	91
25	45	21
26	85	26
27	58	26
28	99	52
29	90	58
30	76	56

D		
1	33	56
2	30	66
3	25	45
4	30	45
5	65	12
6	66	65
7	47	13
8	60	35
9	13	96
10	55	88
11	30	30
12	36	30
13	64	25
14	12	39
15	72	45
16	85	54
17	52	47
18	39	60
19	55	34
20	51	56
21	58	96
22	32	55
23	88	45
24	75	22
25	76	25
26	85	9
27	25	58
28	99	5
29	45	90
30	88	76

WORKBOOK WORK - Reusable - Page 3

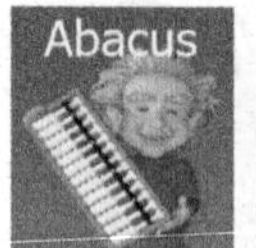

A

1	25	-20
2	25	-13
3	52	-30
4	85	-41
5	25	-9
6	66	-54
7	52	-13
8	39	-20
9	74	-20
10	79	-15
11	95	-65
12	32	-32
13	65	-7
14	12	-8
15	72	-45
16	85	-65
17	25	-22
18	10	-8
19	11	-5
20	51	-40
21	64	-33
22	48	-3
23	85	-2
24	48	-7
25	12	-5
26	88	-41
27	21	-14
28	99	-2
29	47	-12
30	76	-3

B

1	42	-3
2	32	-6
3	45	-8
4	65	-4
5	22	-3
6	37	-15
7	64	-32
8	51	-41
9	64	-50
10	45	-6
11	65	-3
12	64	-23
13	12	-10
14	94	-4
15	55	-47
16	55	-30
17	92	-53
18	75	-23
19	76	-55
20	65	-5
21	85	-70
22	49	-19
23	85	-14
24	91	-5
25	21	-6
26	26	-1
27	26	-12
28	52	-7
29	58	-47
30	56	-26

C

1	66	-33
2	63	-35
3	39	-12
4	45	-8
5	22	-12
6	65	-5
7	44	-41
8	55	-8
9	96	-15
10	85	-9
11	30	-4
12	30	-12
13	25	-16
14	88	-74
15	65	-33
16	98	-66
17	47	-6
18	60	-13
19	13	-2
20	56	-51
21	96	-12
22	55	-44
23	44	-33
24	22	-3
25	45	-2
26	85	-9
27	58	-5
28	99	-41
29	90	-14
30	76	-2

D

1	55	-10
2	20	-8
3	40	-14
4	32	-13
5	82	-9
6	25	-18
7	76	-15
8	33	-14
9	66	-44
10	55	-12
11	30	-16
12	36	-16
13	65	-45
14	36	-12
15	72	-33
16	85	-25
17	25	-12
18	39	-10
19	55	-24
20	51	-8
21	64	-12
22	32	-5
23	92	-41
24	75	-8
25	76	-15
26	85	-9
27	58	-4
28	99	-12
29	90	-23
30	77	-6

WORKBOOK WORK - Reusable - Page 4

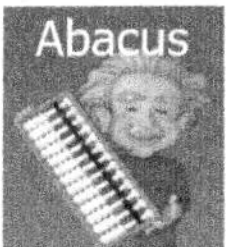

A

1	44	-10
2	35	-8
3	45	-14
4	65	-13
5	25	-9
6	37	-4
7	64	-15
8	44	-14
9	64	-25
10	45	-12
11	66	-16
12	32	-16
13	65	-32
14	12	-12
15	72	-33
16	88	-25
17	25	-12
18	10	-8
19	55	-25
20	51	-8
21	66	-12
22	48	-8
23	85	-41
24	44	-8
25	12	-11
26	88	-9
27	32	-4
28	99	-14
29	47	-23
30	77	-6

B

1	55	-33
2	55	-35
3	77	-12
4	85	-8
5	30	-14
6	44	-5
7	99	-88
8	88	-8
9	65	-15
10	78	-9
11	47	-7
12	64	-12
13	85	-16
14	94	-74
15	55	-21
16	87	-66
17	92	-6
18	75	-13
19	76	-7
20	77	-51
21	85	-15
22	21	-12
23	85	-33
24	91	-3
25	21	-2
26	55	-9
27	26	-8
28	52	-41
29	78	-14
30	56	-4

C

1	66	-3
2	85	-13
3	25	-7
4	64	-4
5	55	-18
6	51	-15
7	77	-32
8	32	-21
9	92	-50
10	37	-6
11	76	-3
12	85	-23
13	58	-10
14	99	-9
15	90	-47
16	88	-30
17	96	-53
18	60	-5
19	65	-55
20	56	-5
21	77	-70
22	55	-35
23	44	-14
24	55	-5
25	45	-7
26	85	-15
27	75	-64
28	80	-7
29	90	-80
30	79	-26

D

1	67	-20
2	55	-26
3	92	-30
4	75	-43
5	84	-9
6	65	-32
7	97	-13
8	33	-25
9	94	-20
10	55	-27
11	99	-65
12	36	-14
13	66	-8
14	36	-9
15	84	-45
16	85	-36
17	25	-22
18	39	-7
19	49	-5
20	51	-31
21	64	-33
22	91	-3
23	92	-74
24	83	-7
25	76	-4
26	73	-42
27	58	-14
28	99	-15
29	90	-75
30	82	-43

WORKBOOK WORK - Reusable - Page 5

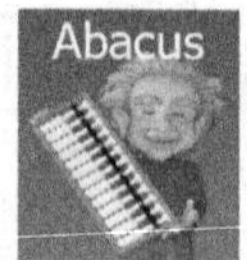

A

1	41	20	-18
2	16	22	-12
3	44	32	-15
4	33	16	-12
5	20	22	-20
6	45	41	-12
7	72	52	-32
8	62	36	-41
9	80	12	-8
10	14	75	-55
11	23	82	-12
12	36	14	-19
13	20	42	-19
14	66	32	-15
15	45	45	-62
16	55	65	-32
17	12	22	-5
18	41	10	-8
19	25	11	-6
20	55	51	-2
21	99	64	-45
22	32	45	-15
23	15	65	-15
24	13	18	-6
25	47	12	-6
26	88	3	-12
27	66	1	-15
28	82	98	-75
29	92	47	-33
30	24	76	-66

B

1	15	85	-5
2	23	15	-12
3	60	45	-15
4	26	35	-8
5	14	23	-4
6	44	45	-12
7	55	74	-12
8	63	62	-25
9	74	82	-9
10	46	14	-54
11	14	38	-13
12	36	33	-20
13	14	20	-20
14	26	65	-16
15	85	45	-65
16	55	54	-32
17	92	12	-7
18	75	12	-8
19	76	13	-3
20	65	55	-2
21	32	96	-45
22	26	32	-14
23	85	12	-33
24	91	13	-4
25	21	47	-2
26	26	85	-9
27	26	66	-5
28	4	82	-42
29	18	91	-14
30	9	23	-2

C

1	20	55	-4
2	25	22	-3
3	32	32	-6
4	16	18	-6
5	22	22	-5
6	44	41	-2
7	52	52	-6
8	36	37	-10
9	12	12	-9
10	79	75	-22
11	19	82	-2
12	32	18	-12
13	24	42	-6
14	12	32	-4
15	72	20	-3
16	32	65	-10
17	64	6	-24
18	10	11	-13
19	11	11	-2
20	51	55	-51
21	64	64	-12
22	48	45	-44
23	65	66	-33
24	18	18	-3
25	12	10	-2
26	32	3	-9
27	35	1	-5
28	98	99	-41
29	47	47	-14
30	76	76	-2

WORKBOOK WORK - Reusable - Page 6

A

1	52	74	-18
2	16	52	-12
3	33	82	-15
4	33	16	-12
5	20	38	-13
6	62	52	-12
7	72	21	-32
8	62	65	-32
9	80	40	-8
10	16	54	-55
11	23	12	-9
12	36	8	-19
13	44	13	-19
14	66	52	-15
15	45	96	-52
16	55	12	-32
17	44	8	-5
18	41	6	-8
19	25	47	-6
20	55	85	-2
21	99	66	-33
22	32	20	-15
23	15	91	-15
24	13	16	-6
25	47	12	-4
26	99	3	-12
27	66	3	-15
28	82	98	-13
29	92	47	-33
30	24	20	-23

B

1	22	85	-5
2	23	82	-12
3	60	12	-16
4	16	12	-8
5	22	47	-4
6	20	85	-9
7	52	16	-8
8	36	88	-25
9	12	91	-9
10	79	23	-44
11	19	17	-13
12	32	9	-20
13	24	9	-18
14	12	28	-12
15	16	47	-15
16	32	76	-5
17	64	22	-20
18	10	55	-12
19	80	13	-32
20	14	30	-41
21	23	96	-6
22	36	32	-55
23	20	11	-12
24	66	11	-4
25	45	47	-16
26	55	55	-9
27	77	66	-25
28	4	82	-42
29	18	36	-14
30	9	39	-2

C

1	10	55	-4
2	10	22	-3
3	32	36	-6
4	16	18	-3
5	22	22	-20
6	44	20	-12
7	64	50	-15
8	20	37	-12
9	11	12	-20
10	51	68	-12
11	64	55	-32
12	95	18	-41
13	85	42	-8
14	91	13	-55
15	21	5	-3
16	26	60	-10
17	28	6	-24
18	4	20	-13
19	11	41	-2
20	60	55	-51
21	64	64	-12
22	48	8	-44
23	65	76	-33
24	18	55	-3
25	99	10	-2
26	32	6	-9
27	35	12	-5
28	98	41	-41
29	47	20	-14
30	76	88	-2

WORKBOOK WORK - Reusable - Page 7

A

1	88	74	-3
2	15	52	-3
3	48	82	-6
4	35	16	-7
5	25	38	-5
6	45	55	-2
7	75	21	-6
8	62	25	-12
9	82	40	-9
10	14	54	-22
11	39	12	-15
12	33	8	-12
13	20	15	-6
14	65	52	-10
15	47	96	-3
16	54	12	-10
17	18	9	-24
18	12	6	-13
19	46	47	-2
20	55	85	-51
21	96	67	-12
22	32	20	-44
23	12	91	-33
24	13	16	-14
25	88	12	-2
26	85	3	-20
27	66	12	-5
28	82	98	-44
29	85	47	-14
30	23	26	-12

B

1	66	85	-15
2	85	88	-12
3	25	12	-15
4	65	12	-18
5	55	47	-20
6	51	95	-13
7	85	16	-20
8	32	88	-41
9	92	99	-7
10	37	23	-45
11	88	17	-13
12	85	15	-15
13	58	9	-22
14	99	36	-15
15	90	47	-65
16	99	76	-13
17	96	22	-7
18	60	56	-6
19	65	15	-9
20	56	30	-4
21	87	96	-26
22	55	32	-12
23	44	15	-15
24	55	11	-7
25	46	47	-4
26	85	60	-12
27	85	66	-16
28	80	82	-44
29	90	37	-33
30	79	42	-55

C

1	55	55	-6
2	30	25	-8
3	25	36	-7
4	33	18	-4
5	65	40	-12
6	66	20	-12
7	47	55	-22
8	60	37	-42
9	45	12	-30
10	55	72	-6
11	30	55	-3
12	36	18	-33
13	64	42	-10
14	24	13	-6
15	72	5	-35
16	85	60	-30
17	55	6	-42
18	39	25	-23
19	55	41	-55
20	51	55	-42
21	65	64	-12
22	32	28	-5
23	88	76	-72
24	85	55	-18
25	76	12	-12
26	85	6	-15
27	25	12	-9
28	88	52	-9
29	45	20	-7
30	75	88	-8

WORKBOOK WORK - Reusable - Page 8

A

1	10	20	-15	-5
2	30	22	-12	-3
3	40	32	-15	-2
4	5	16	-14	-4
5	10	22	-20	-5
6	30	41	-13	-1
7	10	52	-30	-5
8	30	36	-41	-10
9	10	12	-9	-9
10	30	78	-54	-20
11	10	19	-13	-2
12	30	32	-20	-12
13	10	24	-20	-5
14	30	6	-15	-4
15	10	72	-65	-3
16	30	32	-32	-17
17	42	6	-7	-24
18	30	10	-8	-13
19	10	11	-3	-4
20	30	51	-2	-51
21	10	64	-45	-12
22	30	45	-12	-40
23	10	65	-15	-33
24	30	18	-6	-3
25	10	12	-4	-2
26	30	3	-12	-7
27	47	1	-15	-5
28	30	98	-74	-41
29	10	47	-33	-14
30	30	76	-66	-2

B

1	15	-3	63	-5
2	22	-6	15	-3
3	60	-8	42	-2
4	26	-4	35	-4
5	12	-3	27	-5
6	44	-12	45	-1
7	52	-32	65	-5
8	63	-41	75	-10
9	74	-50	82	-9
10	21	-5	14	-20
11	12	-3	38	-2
12	36	-23	32	-12
13	14	-10	20	-5
14	25	-7	65	-4
15	85	-47	41	-3
16	52	-30	54	-17
17	92	-53	12	-24
18	74	-23	9	-13
19	76	-55	13	-4
20	65	-42	54	-51
21	32	-14	96	-12
22	25	-6	32	-40
23	85	-70	9	-6
24	91	-19	2	-3
25	21	-14	47	-2
26	25	-3	85	-7
27	26	-6	61	-5
28	2	-1	82	-41
29	18	-14	90	-14
30	9	-7	23	-2

C

1	20	20	-10	-5
2	65	22	-8	-12
3	41	32	-13	-15
4	54	16	-14	-6
5	12	22	-7	-4
6	9	41	-8	-12
7	13	52	-3	-15
8	54	36	-2	-74
9	96	12	-45	-9
10	30	75	-12	-54
11	10	82	-15	-13
12	30	14	-6	-20
13	10	42	-4	-20
14	30	32	-12	-15
15	65	20	-32	-45
16	54	65	-35	-32
17	45	6	-12	-7
18	60	10	-7	-8
19	13	11	-12	-3
20	54	51	-5	-2
21	96	64	-41	-45
22	32	45	-8	-12
23	9	65	-15	-33
24	2	18	-6	-3
25	47	12	-4	-2
26	85	3	-12	-7
27	61	1	-15	-5
28	82	98	-74	-41
29	90	47	-33	-14
30	23	76	-66	-2

WORKBOOK WORK - Reusable - Page 9

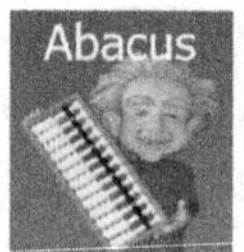

A

1	85	20	-15	-5
2	15	22	-12	-3
3	45	32	-15	-2
4	35	16	-14	-4
5	23	22	-20	-5
6	45	41	-13	-1
7	74	52	-30	-5
8	62	36	-41	-10
9	82	12	-9	-9
10	14	75	-54	-20
11	38	82	-13	-2
12	33	14	-20	-12
13	20	42	-20	-5
14	65	32	-15	-4
15	45	45	-65	-3
16	54	65	-32	-17
17	12	22	-7	-24
18	12	10	-8	-13
19	13	11	-3	-4
20	55	51	-2	-51
21	96	64	-45	-12
22	32	45	-12	-40
23	12	65	-15	-33
24	13	18	-6	-3
25	47	12	-4	-2
26	85	3	-12	-7
27	66	1	-15	-5
28	82	98	-74	-41
29	91	47	-33	-14
30	23	76	-66	-2

B

1	15	-3	20	-5
2	23	-6	25	-12
3	60	-8	32	-15
4	26	-4	16	-8
5	14	-3	22	-4
6	44	-15	44	-12
7	55	-32	52	-12
8	63	-41	36	-25
9	74	-50	12	-9
10	21	-6	79	-54
11	14	-3	19	-13
12	36	-23	32	-20
13	14	-10	24	-20
14	26	-4	12	-16
15	85	-47	72	-65
16	55	-30	32	-32
17	92	-53	6	-7
18	75	-23	10	-8
19	76	-55	11	-3
20	65	-42	51	-2
21	32	-14	64	-45
22	26	-5	48	-14
23	85	-70	65	-33
24	91	-19	18	-4
25	21	-14	12	-2
26	26	-5	3	-9
27	26	-6	1	-5
28	4	-1	98	-42
29	18	-12	47	-14
30	9	-7	76	-2

C

1	21	55	-10	-4
2	66	22	-8	-3
3	41	32	-14	-6
4	45	18	-14	-6
5	12	22	-9	-5
6	9	41	-8	-2
7	13	52	-3	-6
8	55	37	-2	-10
9	96	12	-44	-9
10	31	75	-12	-22
11	10	82	-16	-2
12	30	18	-6	-12
13	12	42	-4	-6
14	30	32	-12	-4
15	65	20	-33	-3
16	54	65	-35	-10
17	47	6	-12	-24
18	60	11	-8	-13
19	13	11	-12	-2
20	54	55	-5	-51
21	96	64	-41	-12
22	32	45	-8	-44
23	11	66	-15	-33
24	2	18	-9	-3
25	45	10	-4	-2
26	85	3	-12	-9
27	58	1	-16	-5
28	82	99	-74	-41
29	90	47	-33	-14
30	25	76	-66	-2

WORKBOOK WORK - Reusable - Page 10

A

1	30	20	-15	-4
2	65	22	-15	-3
3	41	25	-15	-5
4	54	16	-23	-4
5	14	22	-20	-5
6	9	44	-14	-1
7	16	52	-30	-7
8	54	74	-39	-10
9	99	12	-9	-9
10	30	77	-54	-20
11	12	82	-13	-12
12	30	15	-21	-12
13	10	42	-20	-5
14	66	32	-12	-4
15	65	52	-65	-3
16	54	65	-25	-17
17	52	22	-7	-24
18	60	12	-9	-13
19	13	11	-3	-4
20	54	44	-6	-51
21	78	64	-45	-15
22	60	32	-15	-40
23	9	65	-15	-33
24	2	25	-6	-3
25	52	12	-4	-2
26	85	3	-12	-11
27	61	1	-15	-5
28	82	88	-74	-41
29	99	47	-33	-16
30	23	80	-66	-3

B

1	40	-5	40	-5
2	22	-15	25	-10
3	32	-32	25	-12
4	16	-8	16	-8
5	41	-5	22	-4
6	55	-25	26	-12
7	52	-40	53	-14
8	36	-23	33	-25
9	12	-11	15	-9
10	75	-45	70	-54
11	87	-11	20	-13
12	63	-27	33	-20
13	42	-20	13	-20
14	32	-18	54	-16
15	84	-74	72	-66
16	65	-25	60	-32
17	16	-8	20	-7
18	10	-8	15	-8
19	11	-5	12	-3
20	66	-2	12	-2
21	64	-20	64	-45
22	45	-14	24	-14
23	65	-25	66	-33
24	18	-9	18	-4
25	24	-2	21	-2
26	35	-10	9	-9
27	36	-6	3	-5
28	98	-3	98	-42
29	47	-14	54	-14
30	45	-7	25	-2

C

1	36	55	-4	-15
2	25	32	-3	-20
3	56	32	-6	-15
4	16	19	-6	-13
5	62	22	-20	-20
6	25	42	-12	-13
7	52	52	-17	-30
8	33	52	-12	-32
9	12	35	-20	-9
10	74	75	-10	-54
11	19	88	-32	-13
12	52	18	-42	-20
13	32	42	-8	-20
14	54	32	-45	-15
15	72	20	-3	-12
16	60	55	-10	-32
17	18	18	-24	-7
18	15	11	-13	-8
19	12	11	-2	-3
20	12	65	-51	-2
21	64	25	-12	-45
22	48	45	-40	-12
23	66	66	-33	-9
24	18	27	-3	-6
25	88	10	-2	-4
26	18	3	-5	-12
27	20	8	-5	-15
28	98	99	-32	-74
29	36	47	-14	-51
30	35	25	-2	-33

WORKBOOK WORK - Reusable - Page 11

A

1	88	20	-12	-2	-5
2	17	22	-12	-4	-3
3	44	33	-15	-4	-2
4	30	16	-16	-4	-4
5	30	22	-20	-9	-5
6	44	44	-13	-1	-1
7	77	52	-20	-5	-5
8	63	26	-41	-12	-10
9	82	12	-9	-12	-9
10	15	78	-45	-22	-20
11	33	82	-13	-3	-2
12	35	15	-15	-12	-12
13	25	42	-20	-5	-5
14	66	33	-15	-4	-4
15	44	45	-65	-6	-3
16	50	66	-16	-17	-17
17	40	22	-9	-24	-24
18	30	12	-9	-13	-13
19	78	11	-9	-4	-4
20	90	51	-12	-25	-51
21	95	64	-25	-12	-12
22	67	47	-12	-30	-40
23	46	65	-15	-33	-33
24	28	78	-7	-9	-3
25	82	12	-4	-2	-2
26	65	36	-12	-17	-7
27	46	6	-16	-5	-5
28	95	14	-44	-21	-41
29	52	55	-33	-33	-14
30	32	85	-55	-6	-2

B

1	15	-6	40	-5	14
2	22	-6	25	-12	25
3	60	-7	33	-32	30
4	26	-4	16	-8	16
5	15	-12	22	-4	36
6	44	-15	25	-12	44
7	56	-22	52	-40	52
8	63	-41	33	-25	36
9	74	-30	12	-9	12
10	22	-6	70	-54	79
11	14	-3	19	-11	19
12	35	-23	33	-20	33
13	15	-10	24	-20	24
14	26	-6	54	-16	12
15	88	-47	72	-70	62
16	55	-30	60	-35	33
17	99	-42	6	-8	6
18	75	-23	12	-8	14
19	85	-50	12	-3	11
20	65	-42	12	-2	52
21	32	-12	64	-60	66
22	13	-5	48	-14	44
23	85	-70	66	-30	65
24	91	-18	18	-4	44
25	20	-12	22	-2	12
26	26	-9	9	-10	8
27	36	-9	3	-5	1
28	14	-9	98	-25	44
29	18	-6	25	-14	25
30	10	-8	25	-6	12

WORKBOOK WORK - Reusable - Page 12

A

1	20	20	-4	-2	20
2	65	22	-3	-4	22
3	41	33	-6	-4	32
4	54	16	-3	-4	16
5	12	22	-20	-9	22
6	9	44	-12	-1	41
7	13	52	-15	-5	52
8	54	26	-12	-12	36
9	96	12	-20	-12	12
10	30	78	-12	-22	75
11	10	82	-32	-3	82
12	30	33	-41	-12	14
13	10	42	-8	-5	42
14	30	33	-55	-4	32
15	65	45	-3	-6	20
16	54	66	-10	-17	65
17	45	22	-24	-24	6
18	60	12	-13	-13	10
19	13	11	-2	-4	11
20	54	51	-51	-25	51
21	96	64	-12	-12	64
22	60	47	-44	-30	45
23	9	65	-33	-33	65
24	2	78	-3	-9	18
25	47	12	-2	-2	12
26	85	36	-9	-17	3
27	61	6	-5	-5	1
28	82	14	-41	-21	98
29	90	55	-14	-33	47
30	23	85	-2	-6	76

B

1	41	-15	40	-5	14
2	16	-12	25	-12	25
3	44	-15	33	-32	30
4	33	-14	16	-8	16
5	20	-20	22	-4	36
6	45	-13	25	-26	44
7	72	-30	52	-40	52
8	62	-41	33	-25	36
9	80	-9	12	-9	12
10	64	-54	70	-54	79
11	23	-13	19	-11	19
12	36	-20	33	-38	33
13	36	-20	22	-20	24
14	66	-15	54	-16	12
15	45	-3	72	-70	62
16	55	-32	60	-35	33
17	85	-7	6	-8	6
18	41	-8	12	-8	14
19	25	-3	12	-3	11
20	55	-2	12	-2	52
21	99	-45	64	-60	66
22	32	-12	48	-14	44
23	16	-15	66	-30	65
24	13	-6	18	-4	44
25	47	-4	22	-2	12
26	88	-12	9	-10	8
27	66	-15	3	-5	1
28	82	-74	98	-25	44
29	92	-33	25	-14	25
30	99	-66	25	-6	12

WORKBOOK WORK - Reusable - Page 13

A

1	102	242	-3	-200	20
2	22	12	-6	-4	22
3	123	62	-8	-4	32
4	18	121	-100	-4	16
5	22	6	-3	-9	22
6	441	14	-15	-210	41
7	225	11	-32	-5	52
8	550	111	-41	-221	36
9	12	66	-50	-12	12
10	321	44	-6	-22	75
11	82	65	-3	-3	82
12	18	623	-23	-12	14
13	42	42	-10	-22	42
14	251	33	-4	-4	32
15	20	45	-47	-6	20
16	65	412	-30	-17	65
17	333	22	-53	-120	6
18	412	12	-23	-13	10
19	11	223	-55	-4	11
20	550	51	-400	-25	51
21	213	821	-70	-300	64
22	45	90	-19	-60	45
23	166	65	-121	-33	65
24	18	78	-5	-9	18
25	362	12	-6	-2	12
26	875	36	-418	-17	3
27	365	6	-12	-5	1
28	999	14	-555	-21	98
29	400	55	-47	-33	47
30	76	288	-200	-6	76

B

1	300	-15	40	-10	14
2	22	-12	520	-8	25
3	152	-140	33	-14	30
4	36	-14	160	-13	16
5	220	-20	22	-9	36
6	75	-13	125	-18	44
7	882	-255	52	-15	52
8	140	-41	33	-44	36
9	425	-250	12	-44	12
10	632	-155	70	-120	79
11	120	-13	19	-16	19
12	65	-20	33	-16	33
13	600	-200	22	-310	24
14	100	-15	54	-12	12
15	110	-35	72	-33	62
16	665	-32	60	-150	33
17	452	-200	6	-120	6
18	210	-8	12	-10	14
19	110	-3	130	-24	11
20	230	-25	12	-8	52
21	64	-45	64	-12	66
22	125	-12	233	-155	44
23	610	-250	66	-200	65
24	180	-80	18	-8	44
25	120	-4	22	-15	12
26	125	-25	9	-9	8
27	225	-106	3	-40	1
28	99	-25	188	-12	44
29	850	-450	25	-23	25
30	320	-120	25	-6	12

WORKBOOK WORK - Reusable - Page 14

A

1	120	242	-3	-12	20
2	222	100	-120	-4	22
3	133	62	-8	-25	32
4	16	121	-100	-4	16
5	250	6	-120	-9	22
6	44	320	-15	-120	41
7	520	225	-365	-5	52
8	26	111	-41	-12	36
9	120	66	-50	-120	12
10	780	44	-400	-22	75
11	82	665	-3	-3	82
12	330	623	-23	-12	14
13	42	250	-25	-5	42
14	124	33	-4	-24	32
15	450	45	-200	-6	20
16	66	412	-30	-17	65
17	222	22	-30	-24	6
18	12	444	-23	-44	10
19	110	223	-55	-4	11
20	51	515	-400	-25	51
21	132	821	-70	-520	64
22	165	90	-19	-30	45
23	65	241	-121	-14	65
24	136	78	-5	-9	18
25	12	444	-6	-2	12
26	950	555	-418	-600	3
27	963	6	-12	-540	1
28	652	100	-555	-21	98
29	55	200	-47	-33	47
30	425	288	-200	-6	76

B

1	240	-6	120	-10	14
2	125	-6	420	-8	25
3	130	-7	55	-14	30
4	160	-4	160	-13	16
5	360	-12	140	-145	36
6	440	-15	125	-250	44
7	225	-200	52	-15	52
8	635	-410	33	-44	36
9	127	-30	12	-44	12
10	790	-650	70	-120	79
11	250	-54	19	-16	19
12	330	-23	33	-16	33
13	240	-110	200	-310	24
14	120	-60	54	-12	12
15	665	-470	72	-33	62
16	352	-300	500	-150	33
17	362	-320	600	-120	6
18	145	-23	12	-10	14
19	110	-50	130	-24	11
20	520	-420	12	-8	52
21	660	-500	64	-12	66
22	415	-350	233	-155	44
23	254	-145	66	-20	65
24	400	-200	18	-8	44
25	120	-25	22	-15	12
26	880	-741	9	-9	8
27	199	-99	3	-40	1
28	420	-145	188	-12	44
29	250	-125	25	-23	25
30	150	-50	25	-6	12

ANSWERS - 1

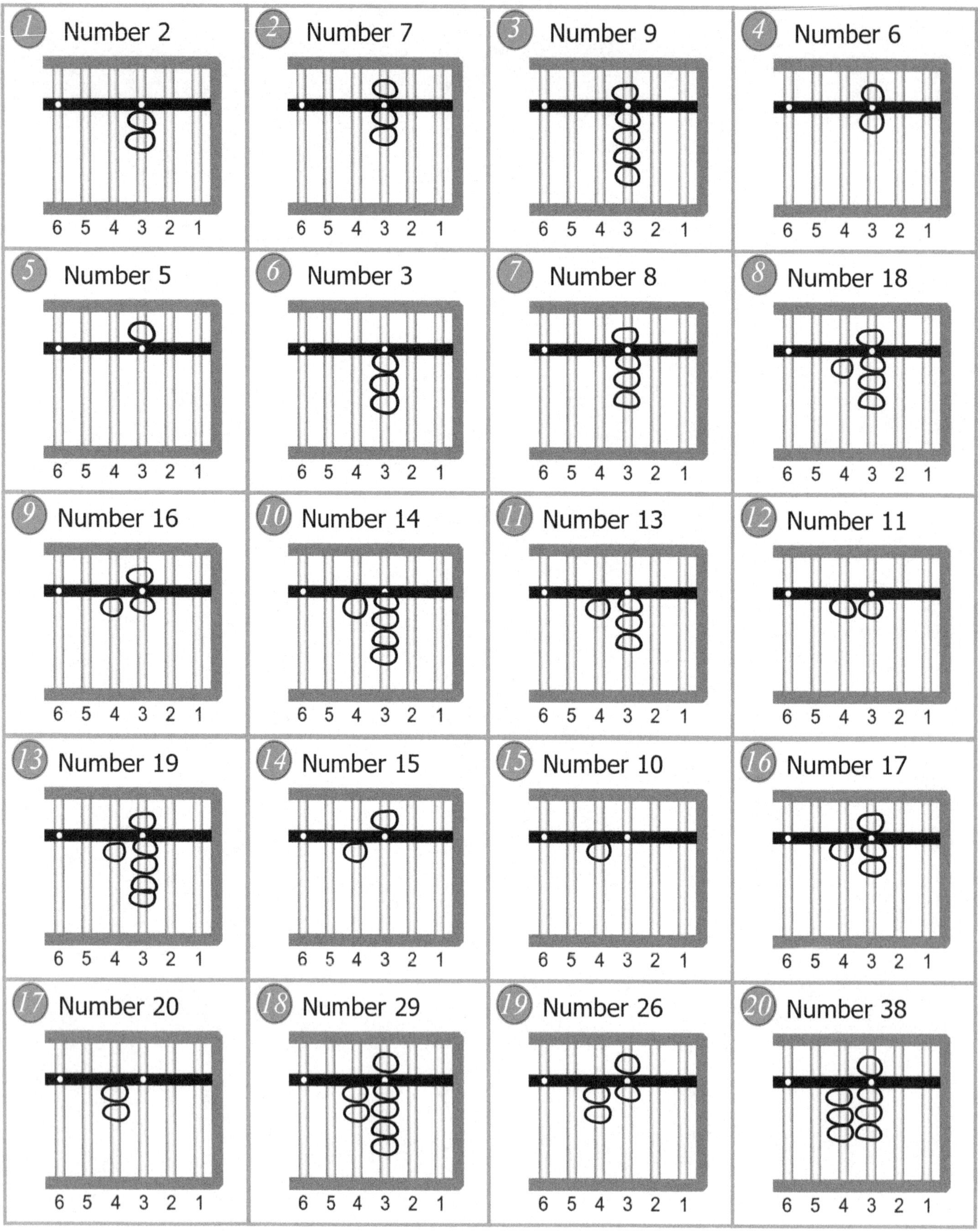

1 Number 2
6 5 4 3 2 1
2 Number 7
6 5 4 3 2 1
3 Number 9
6 5 4 3 2 1
4 Number 6
6 5 4 3 2 1
5 Number 5
6 5 4 3 2 1
6 Number 3
6 5 4 3 2 1
7 Number 8
6 5 4 3 2 1
8 Number 18
6 5 4 3 2 1
9 Number 16
6 5 4 3 2 1
10 Number 14
6 5 4 3 2 1
11 Number 13
6 5 4 3 2 1
12 Number 11
6 5 4 3 2 1
13 Number 19
6 5 4 3 2 1
14 Number 15
6 5 4 3 2 1
15 Number 10
6 5 4 3 2 1
16 Number 17
6 5 4 3 2 1
17 Number 20
6 5 4 3 2 1
18 Number 29
6 5 4 3 2 1
19 Number 26
6 5 4 3 2 1
20 Number 38
6 5 4 3 2 1

ANSWERS - 1

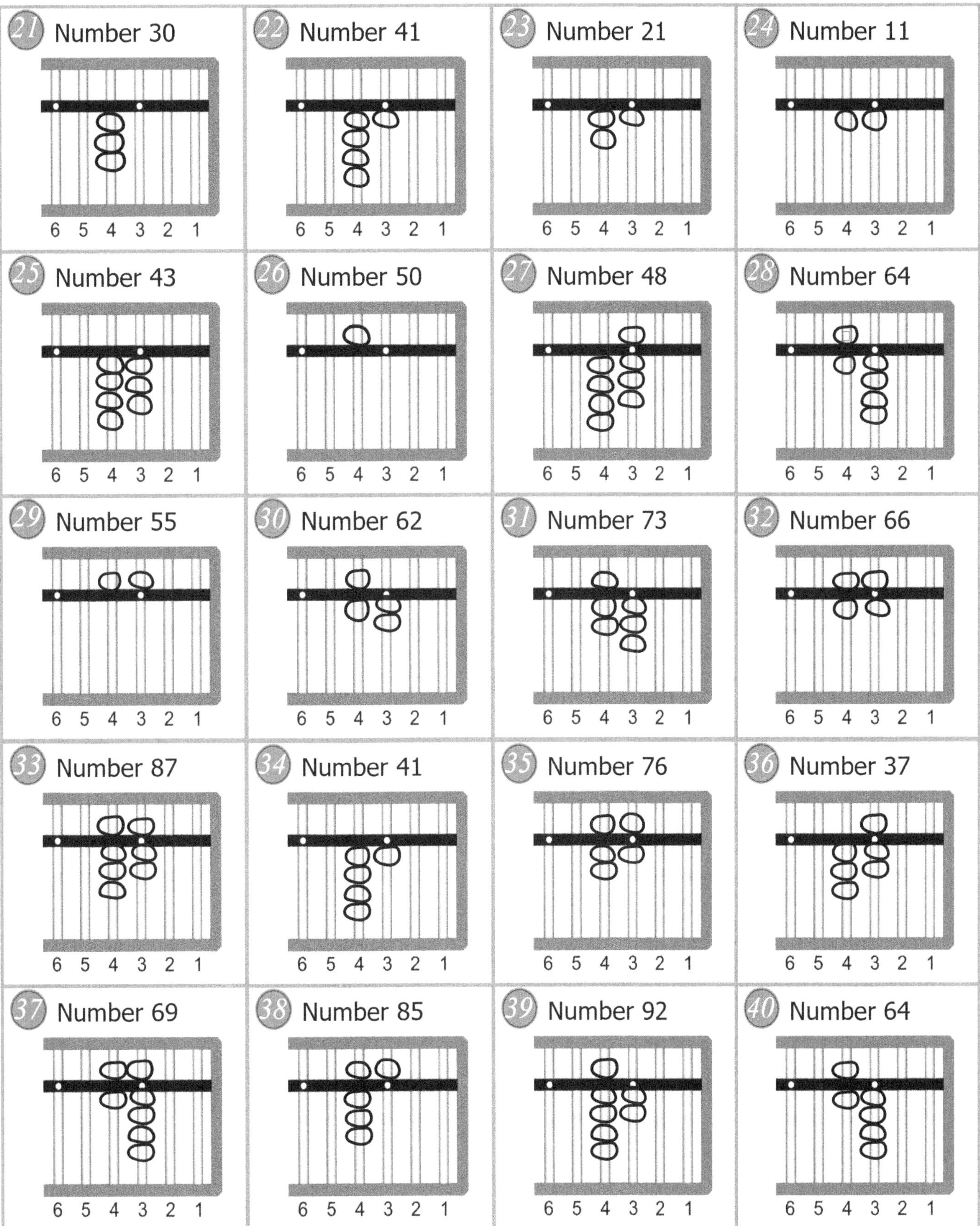

21 Number 30
22 Number 41
23 Number 21
24 Number 11
25 Number 43
26 Number 50
27 Number 48
28 Number 64
29 Number 55
30 Number 62
31 Number 73
32 Number 66
33 Number 87
34 Number 41
35 Number 76
36 Number 37
37 Number 69
38 Number 85
39 Number 92
40 Number 64
6 5 4 3 2 1

ANSWERS - 1

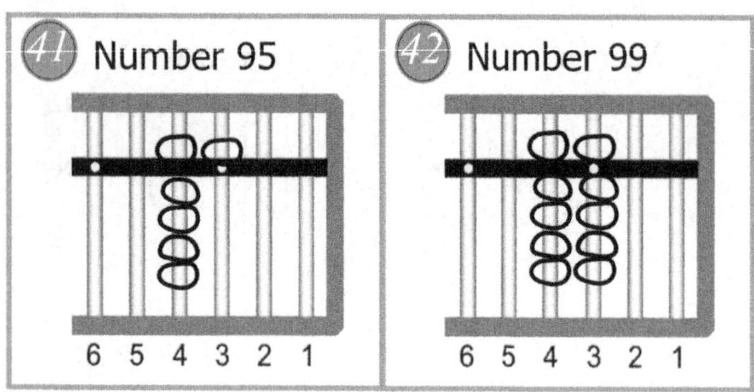

43 Digit 8 of number 8 — 6 5 4 (3) 2 1

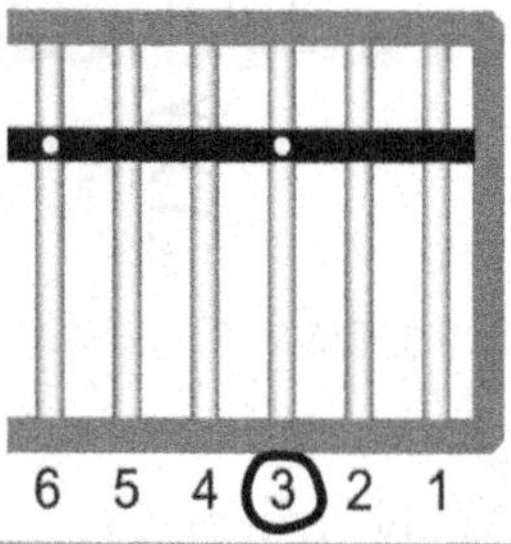

44 Digit 5 of number 35 — 6 5 4 (3) 2 1

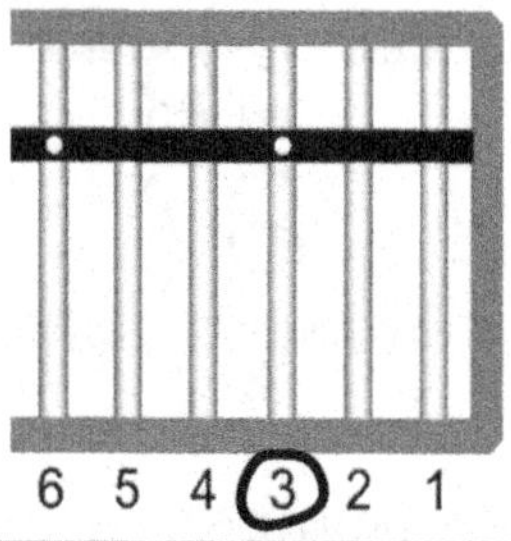

45 Digit 3 of number 35 — 6 5 (4) 3 2 1

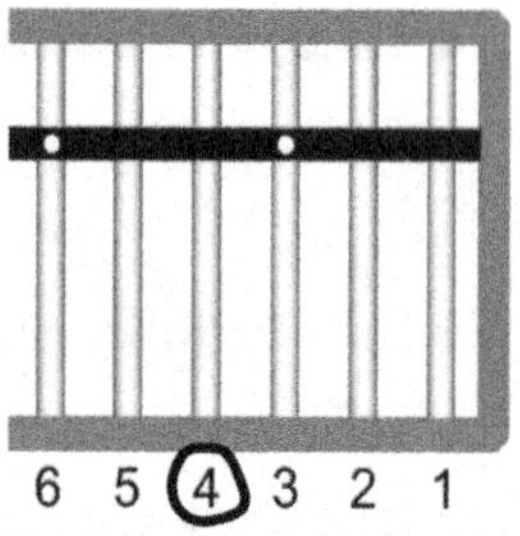

46 Digit 2 of number 28 — 6 5 (4) 3 2 1

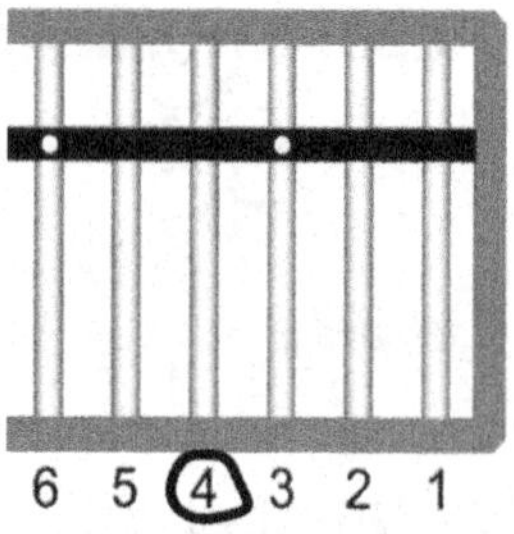

47 Digit 6 of number 65 — 6 5 (4) 3 2 1

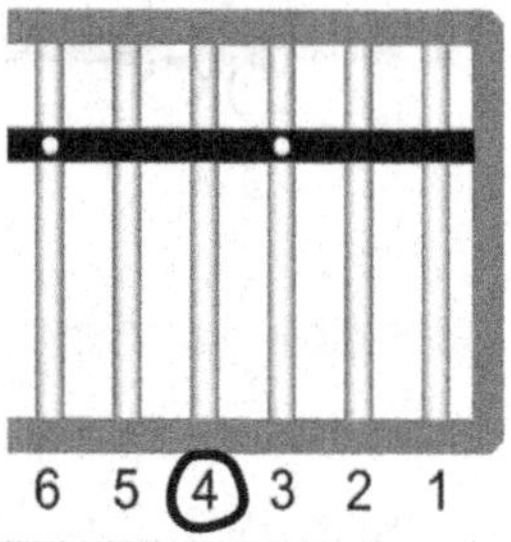

48 Digit 8 of number 78 — 6 5 4 (3) 2 1

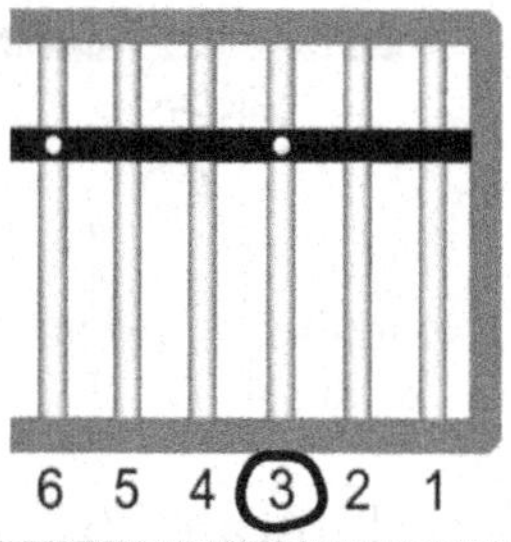

49 Digit 7 of number 27 — 6 5 4 (3) 2 1

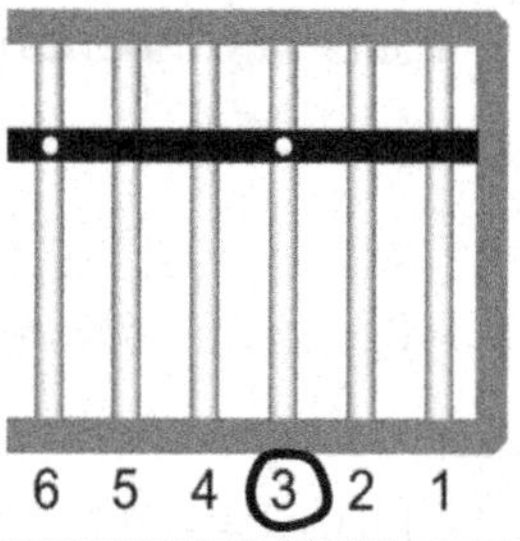

50 Digit 8 of number 48 — 6 5 4 (3) 2 1

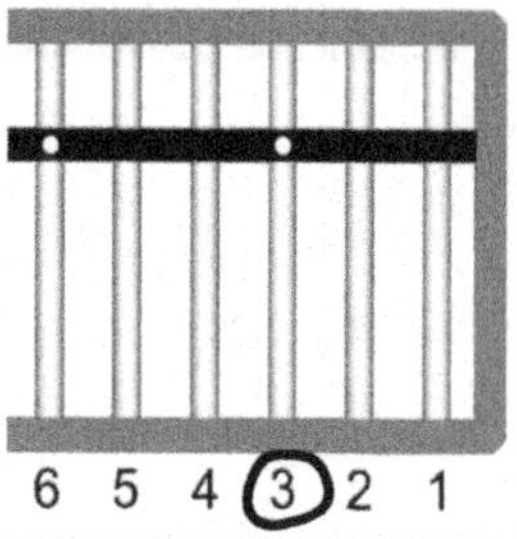

51 Digit 5 of number 56 — 6 5 (4) 3 2 1

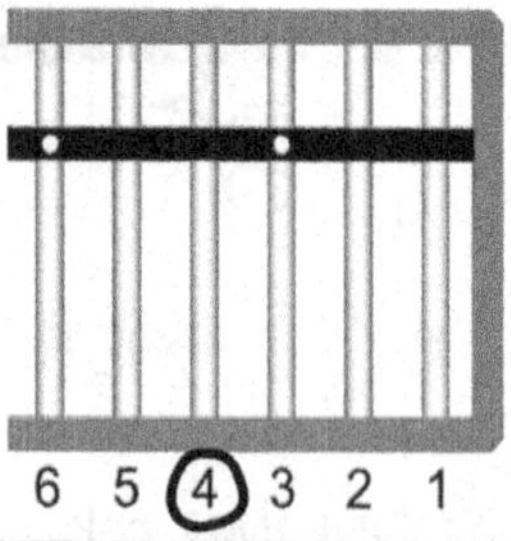

52 Digit 6 of number 61 — 6 5 (4) 3 2 1

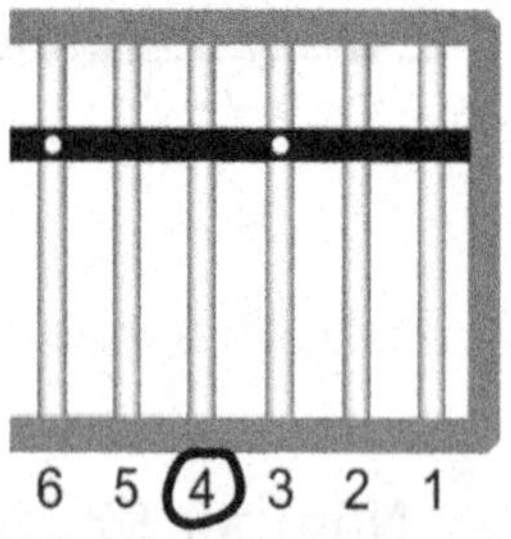

53 Digit 3 of number 83 — 6 5 4 (3) 2 1

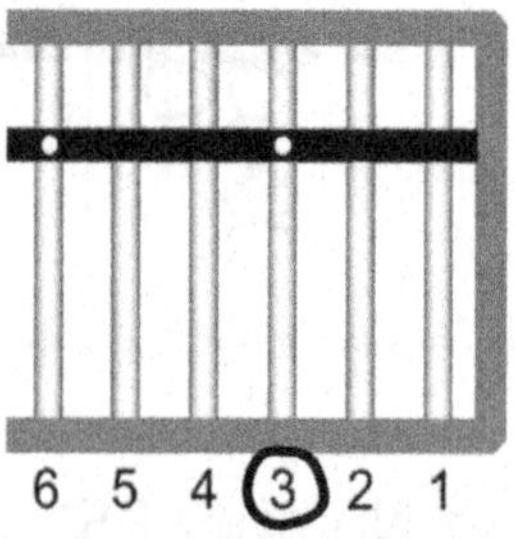

54 Digit 4 of number 42 — 6 5 (4) 3 2 1

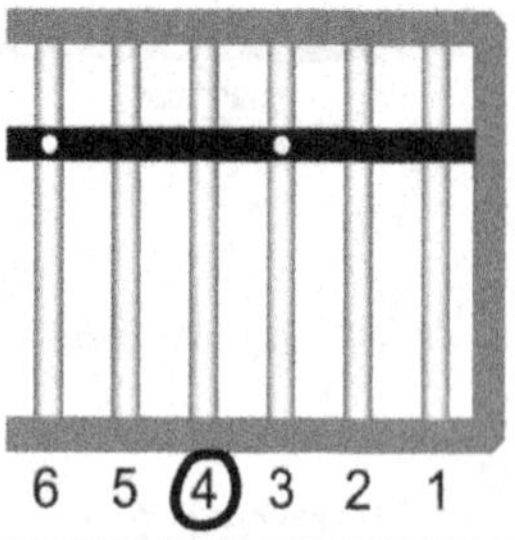

55 Digit 1 of number 21 — 6 5 4 (3) 2 1

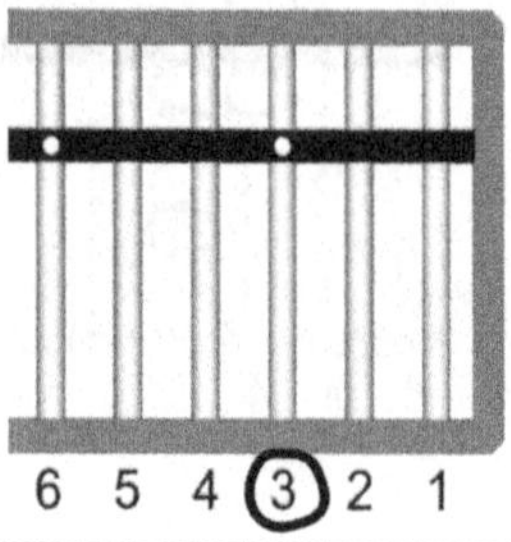

56 Digit 1 of number 15 — 6 5 (4) 3 2 1

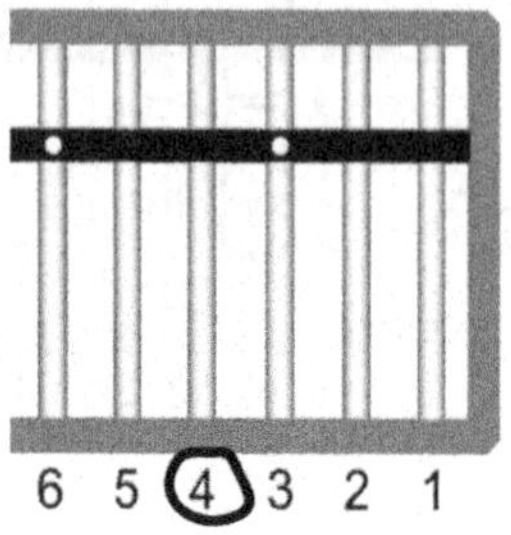

57 Digit 9 of number 94 — 6 5 (4) 3 2 1

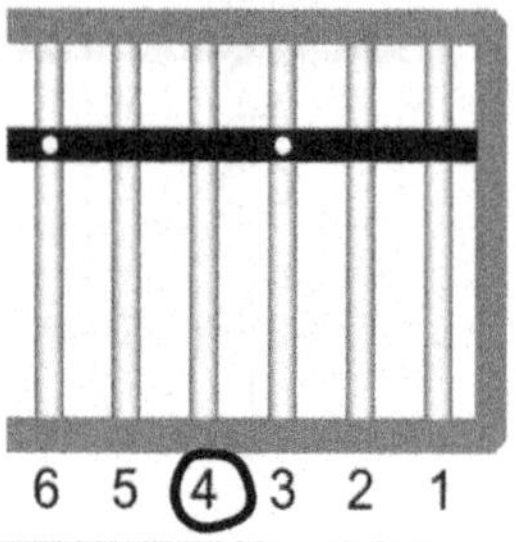

ANSWERS - 2

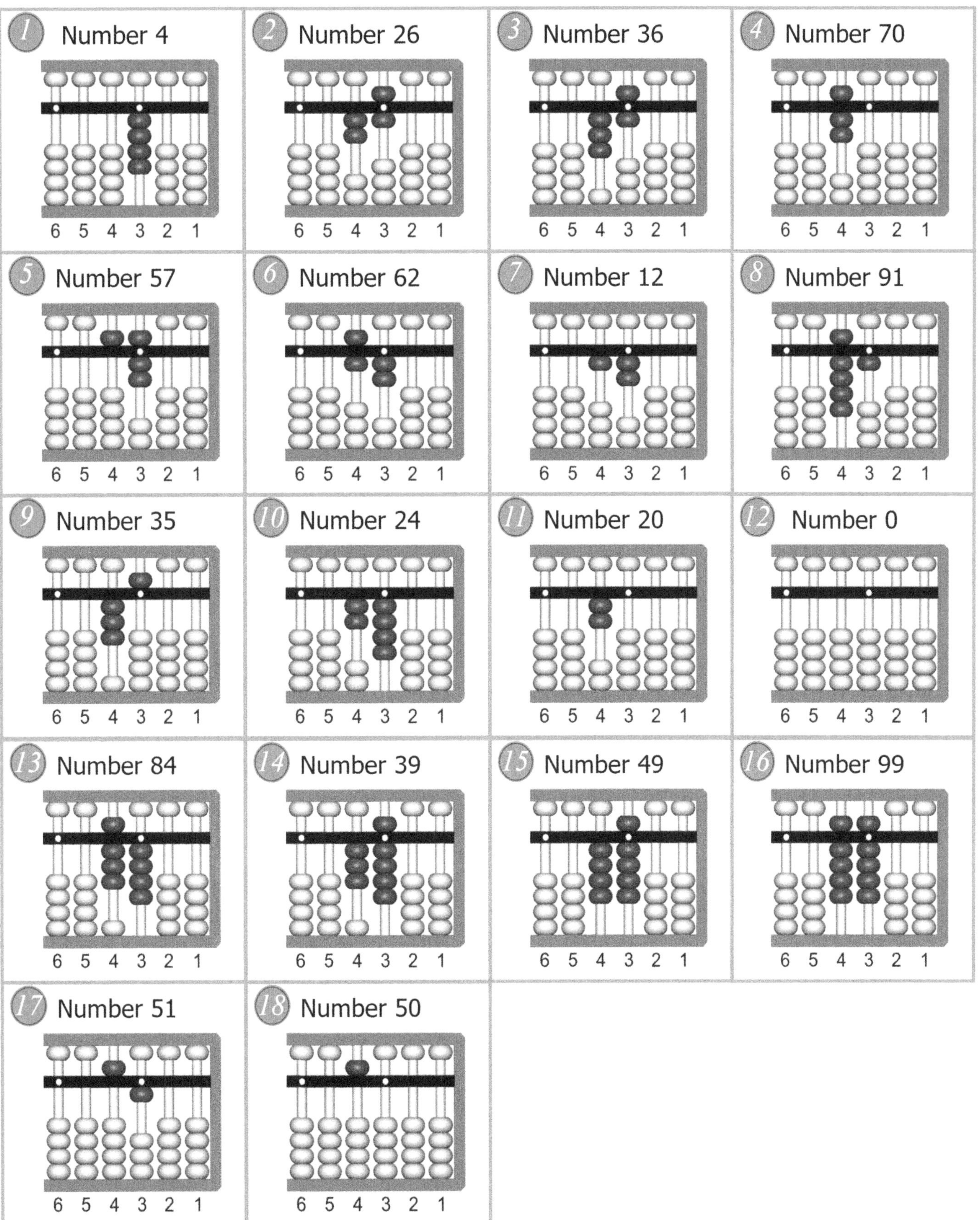
1 Number 4
6 5 4 3 2 1
2 Number 26
6 5 4 3 2 1
3 Number 36
6 5 4 3 2 1
4 Number 70
6 5 4 3 2 1
5 Number 57
6 5 4 3 2 1
6 Number 62
6 5 4 3 2 1
7 Number 12
6 5 4 3 2 1
8 Number 91
6 5 4 3 2 1
9 Number 35
6 5 4 3 2 1
10 Number 24
6 5 4 3 2 1
11 Number 20
6 5 4 3 2 1
12 Number 0
6 5 4 3 2 1
13 Number 84
6 5 4 3 2 1
14 Number 39
6 5 4 3 2 1
15 Number 49
6 5 4 3 2 1
16 Number 99
6 5 4 3 2 1
17 Number 51
6 5 4 3 2 1
18 Number 50
6 5 4 3 2 1

ANSWERS - 3

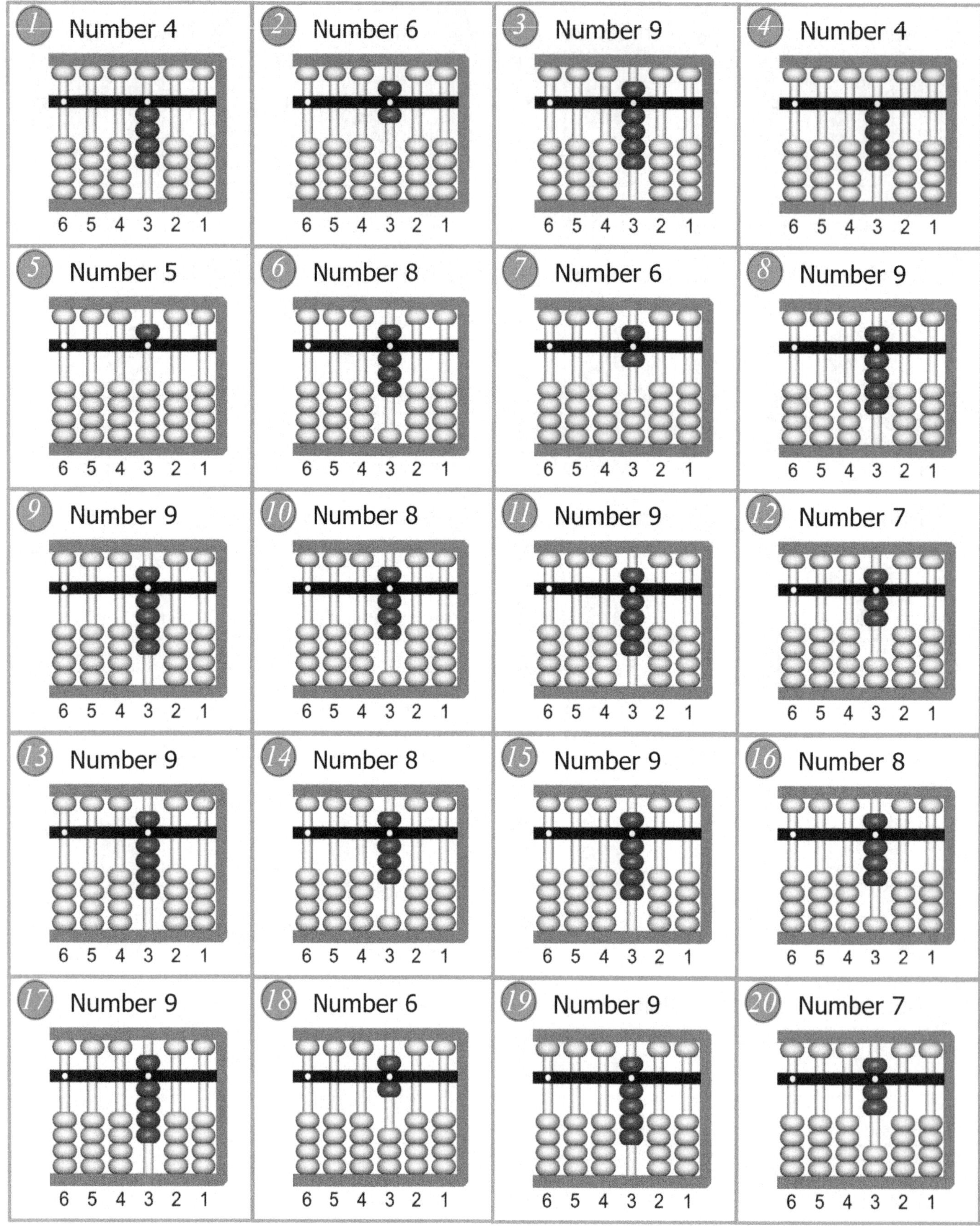
1 Number 4
6 5 4 3 2 1
2 Number 6
6 5 4 3 2 1
3 Number 9
6 5 4 3 2 1
4 Number 4
6 5 4 3 2 1
5 Number 5
6 5 4 3 2 1
6 Number 8
6 5 4 3 2 1
7 Number 6
6 5 4 3 2 1
8 Number 9
6 5 4 3 2 1
9 Number 9
6 5 4 3 2 1
10 Number 8
6 5 4 3 2 1
11 Number 9
6 5 4 3 2 1
12 Number 7
6 5 4 3 2 1
13 Number 9
6 5 4 3 2 1
14 Number 8
6 5 4 3 2 1
15 Number 9
6 5 4 3 2 1
16 Number 8
6 5 4 3 2 1
17 Number 9
6 5 4 3 2 1
18 Number 6
6 5 4 3 2 1
19 Number 9
6 5 4 3 2 1
20 Number 7
6 5 4 3 2 1

ANSWERS - 3

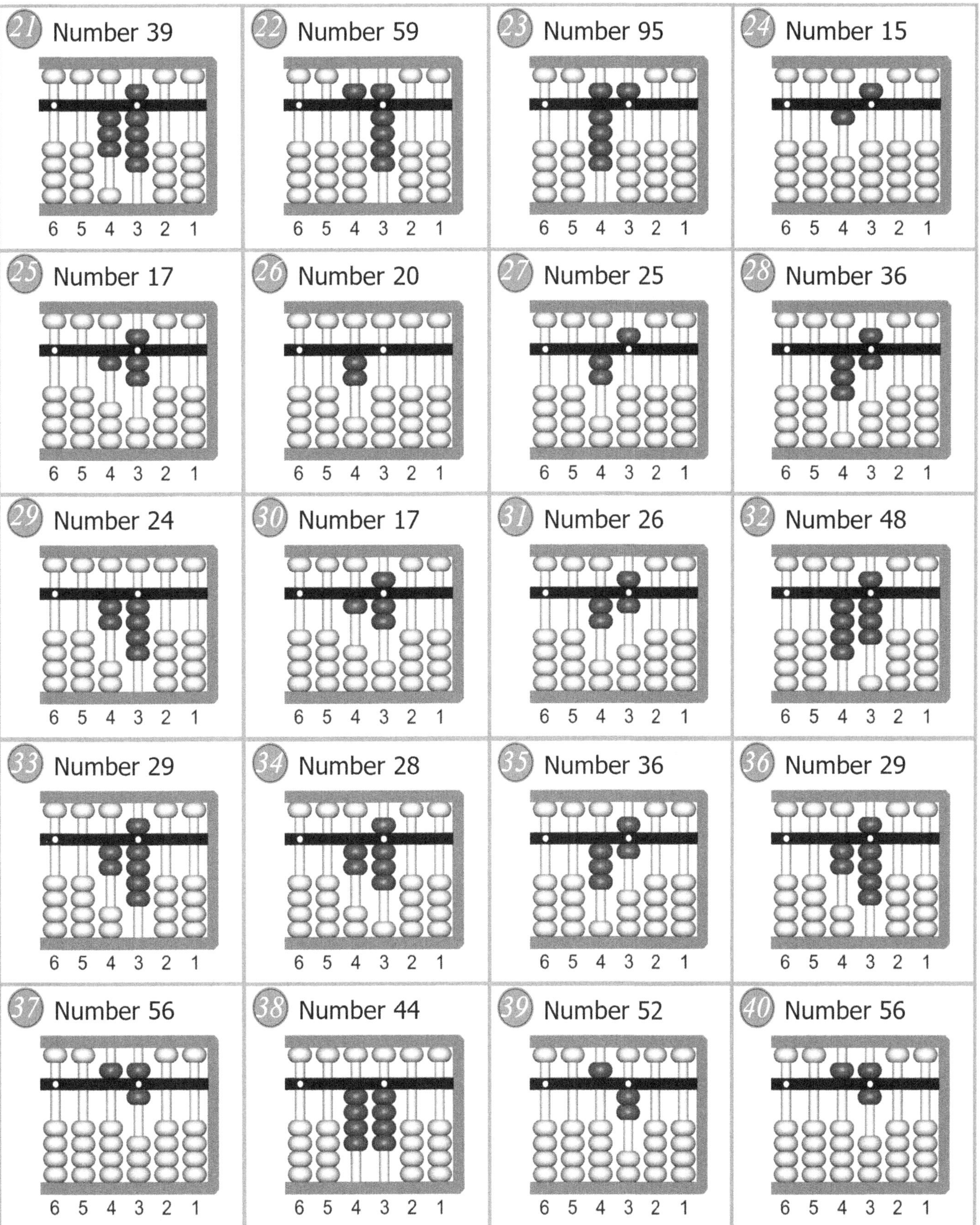
21 Number 39
6 5 4 3 2 1
22 Number 59
6 5 4 3 2 1
23 Number 95
6 5 4 3 2 1
24 Number 15
6 5 4 3 2 1
25 Number 17
6 5 4 3 2 1
26 Number 20
6 5 4 3 2 1
27 Number 25
6 5 4 3 2 1
28 Number 36
6 5 4 3 2 1
29 Number 24
6 5 4 3 2 1
30 Number 17
6 5 4 3 2 1
31 Number 26
6 5 4 3 2 1
32 Number 48
6 5 4 3 2 1
33 Number 29
6 5 4 3 2 1
34 Number 28
6 5 4 3 2 1
35 Number 36
6 5 4 3 2 1
36 Number 29
6 5 4 3 2 1
37 Number 56
6 5 4 3 2 1
38 Number 44
6 5 4 3 2 1
39 Number 52
6 5 4 3 2 1
40 Number 56
6 5 4 3 2 1

ANSWERS - 3

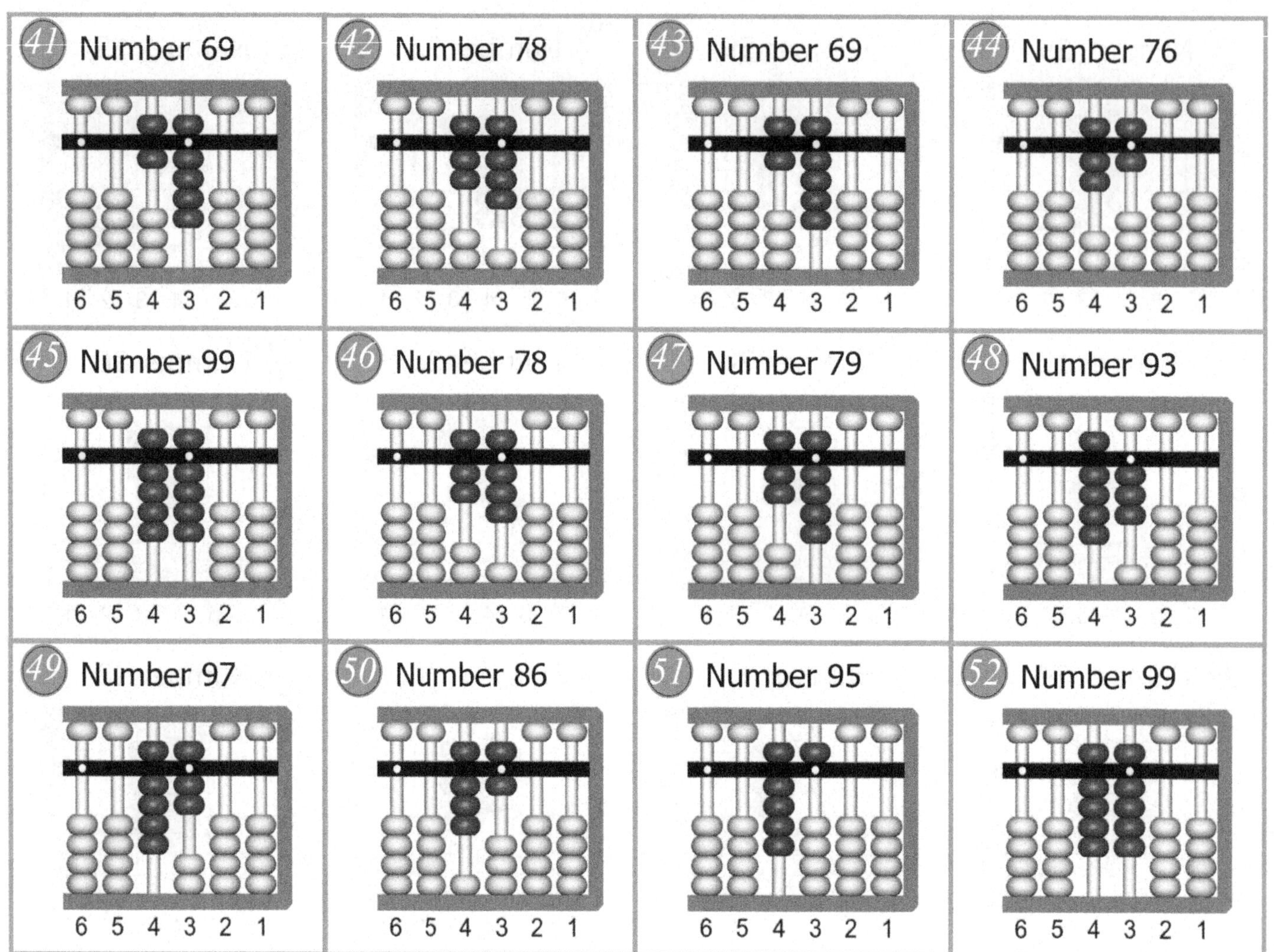
41 Number 69
6 5 4 3 2 1
42 Number 78
6 5 4 3 2 1
43 Number 69
6 5 4 3 2 1
44 Number 76
6 5 4 3 2 1
45 Number 99
6 5 4 3 2 1
46 Number 78
6 5 4 3 2 1
47 Number 79
6 5 4 3 2 1
48 Number 93
6 5 4 3 2 1
49 Number 97
6 5 4 3 2 1
50 Number 86
6 5 4 3 2 1
51 Number 95
6 5 4 3 2 1
52 Number 99
6 5 4 3 2 1

ANSWERS - 3

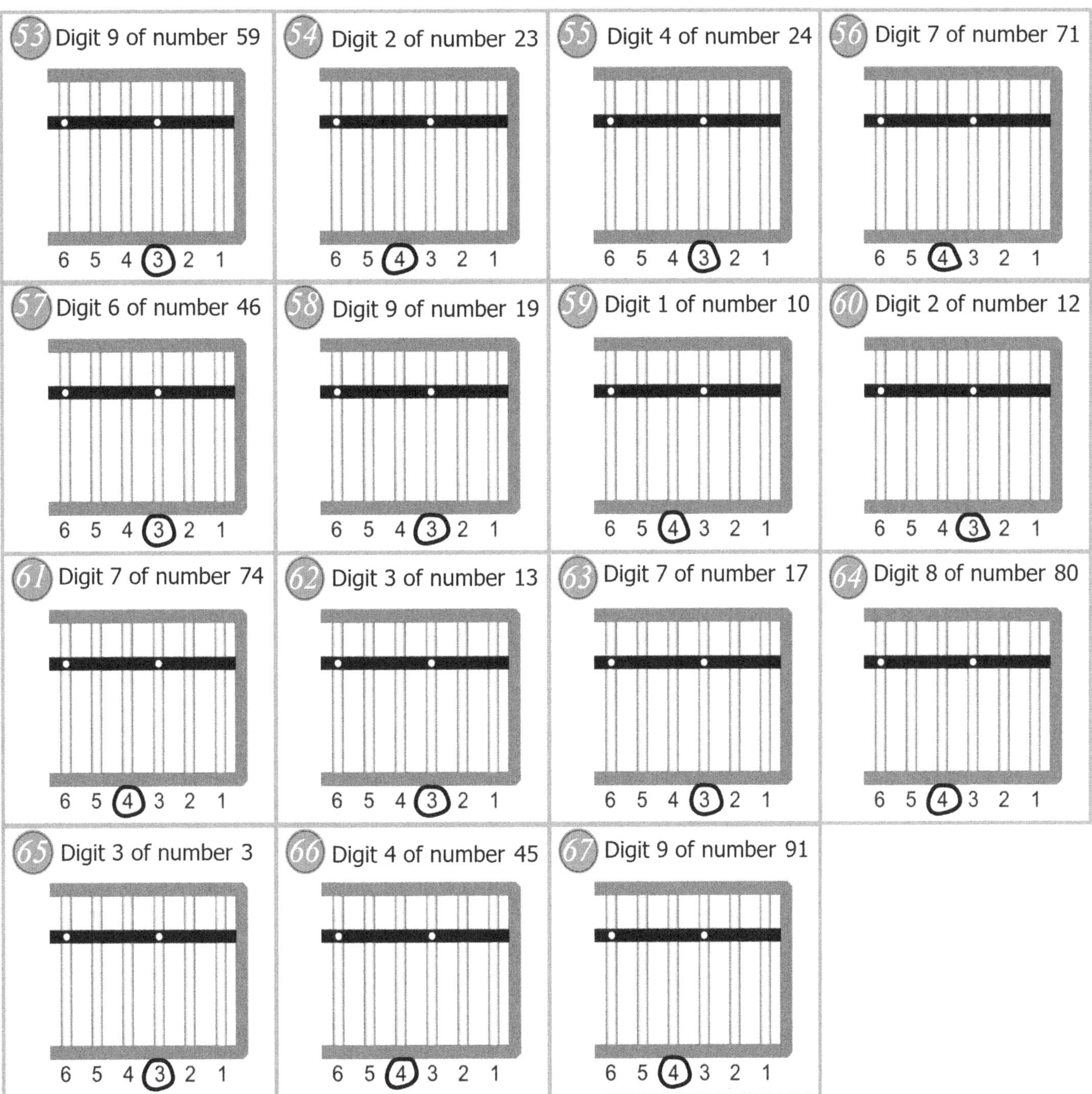

53 Digit 9 of number 59
6 5 4 (3) 2 1
54 Digit 2 of number 23
6 5 (4) 3 2 1
55 Digit 4 of number 24
6 5 4 (3) 2 1
56 Digit 7 of number 71
6 5 (4) 3 2 1
57 Digit 6 of number 46
6 5 4 (3) 2 1
58 Digit 9 of number 19
6 5 4 (3) 2 1
59 Digit 1 of number 10
6 5 (4) 3 2 1
60 Digit 2 of number 12
6 5 4 (3) 2 1
61 Digit 7 of number 74
6 5 (4) 3 2 1
62 Digit 3 of number 13
6 5 4 (3) 2 1
63 Digit 7 of number 17
6 5 4 (3) 2 1
64 Digit 8 of number 80
6 5 (4) 3 2 1
65 Digit 3 of number 3
6 5 4 (3) 2 1
66 Digit 4 of number 45
6 5 (4) 3 2 1
67 Digit 9 of number 91
6 5 (4) 3 2 1

ANSWERS - 4

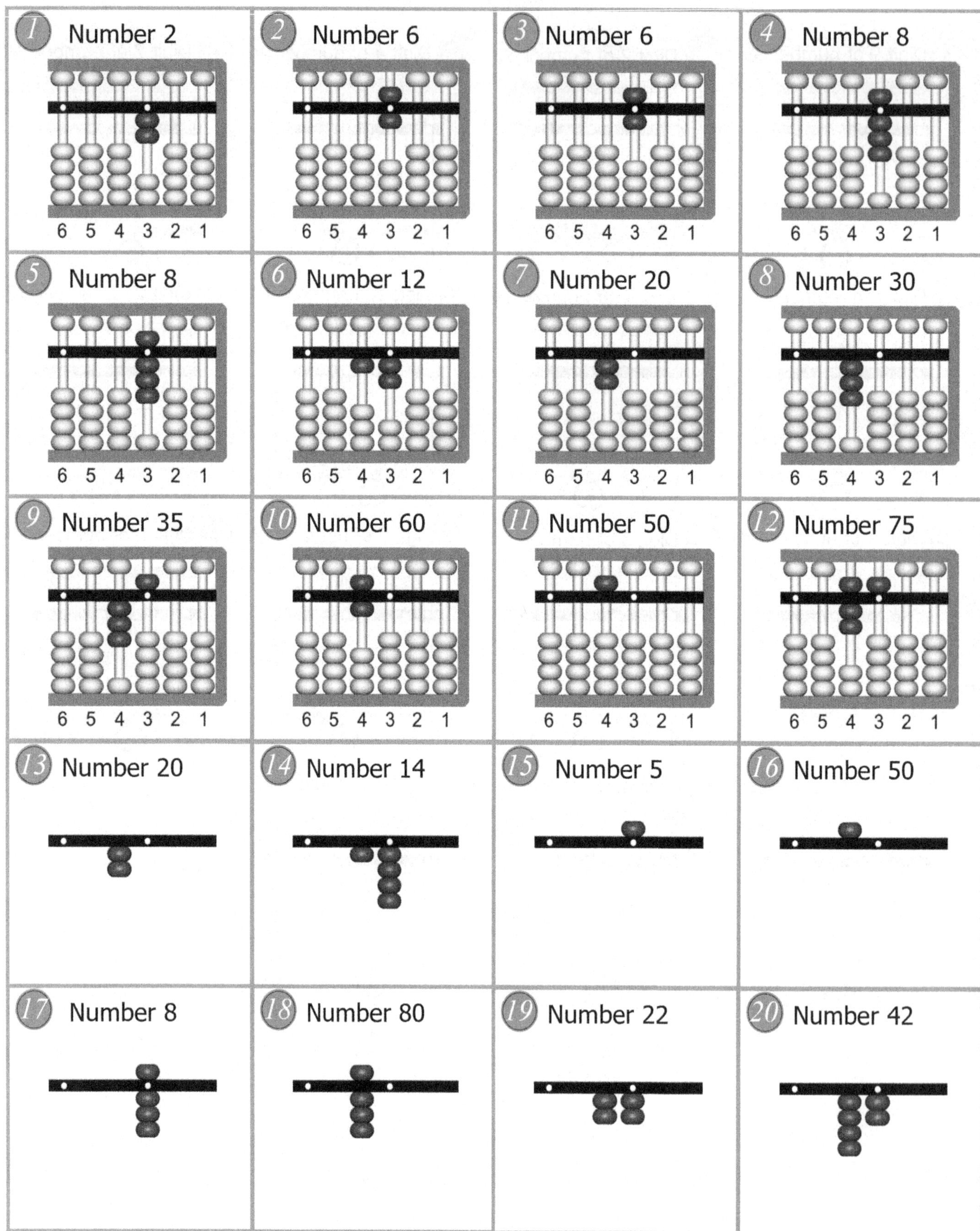
1 Number 2
6 5 4 3 2 1
2 Number 6
6 5 4 3 2 1
3 Number 6
6 5 4 3 2 1
4 Number 8
6 5 4 3 2 1
5 Number 8
6 5 4 3 2 1
6 Number 12
6 5 4 3 2 1
7 Number 20
6 5 4 3 2 1
8 Number 30
6 5 4 3 2 1
9 Number 35
6 5 4 3 2 1
10 Number 60
6 5 4 3 2 1
11 Number 50
6 5 4 3 2 1
12 Number 75
6 5 4 3 2 1
13 Number 20
14 Number 14
15 Number 5
16 Number 50
17 Number 8
18 Number 80
19 Number 22
20 Number 42

ANSWERS - 4

21 Number 49	22 Number 16	23 Number 0	24 Number 87
25 Number 10	26 Number 25	27 Number 41	28 Number 9
29 Number 74	30 Number 12	31 Number 22	32 Number 78
33 Number 14	34 Number 95	35 Number 58	36 Number 66
37 Number 4	38 Number 55	39 Number 92	40 Number 2

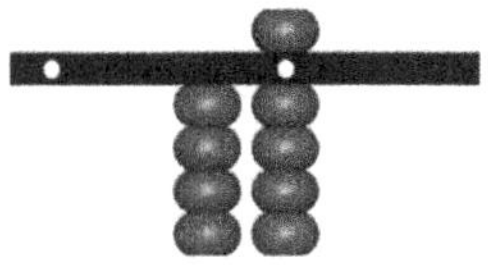

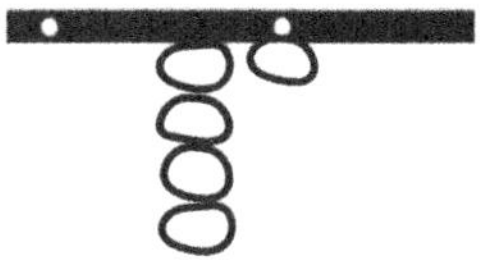

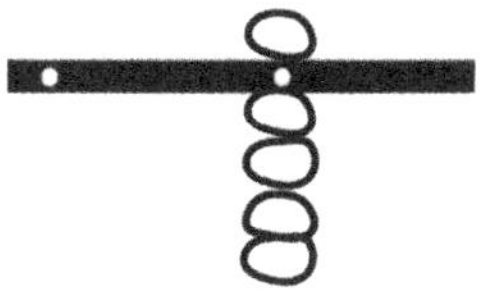

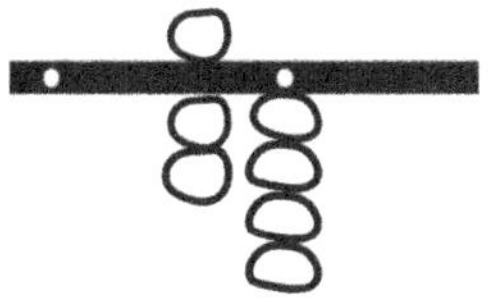

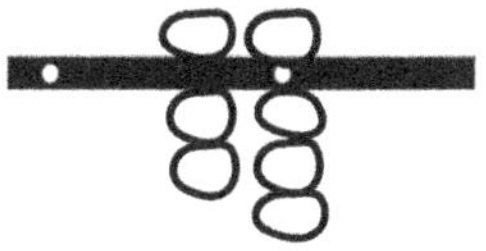

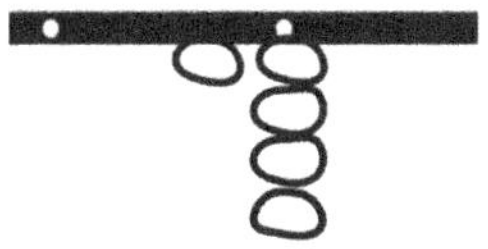

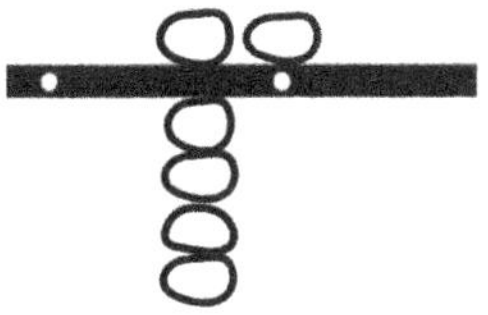

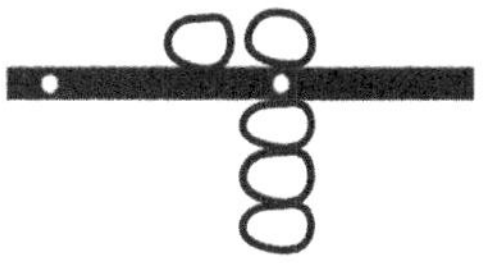

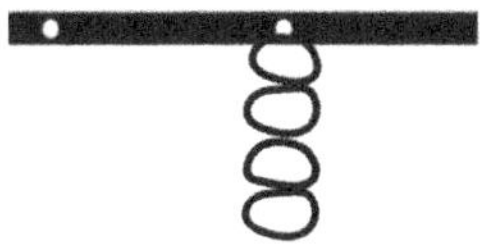

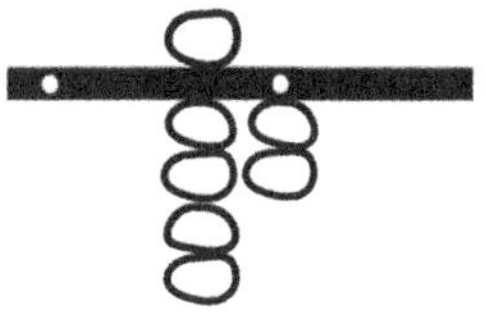

ANSWERS - 4

41 Number 20 	42 Number 33 	43 Number 55 	44 Number 63 
45 Number 15 	46 Number 7 	47 Number 36 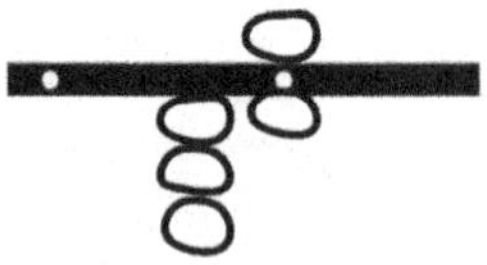	48 Number 39 
49 Number 94 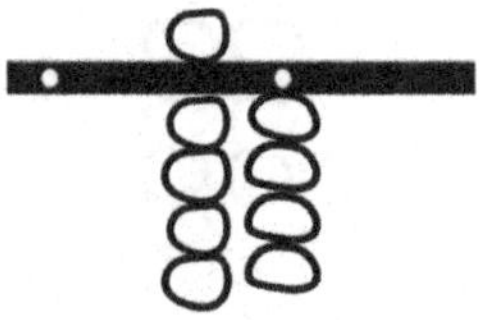	50 Number 57 	51 Number 27 	52 Number 82 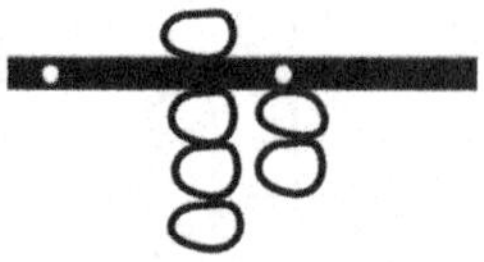
53 Number 13 	54 Number 88 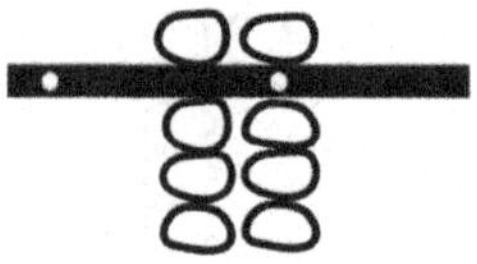	55 Number 4 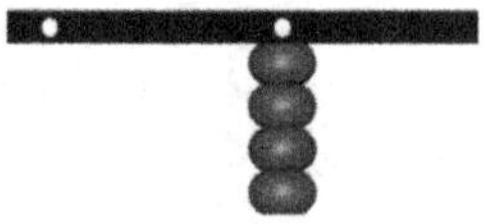	56 Number 6 
57 Number 9 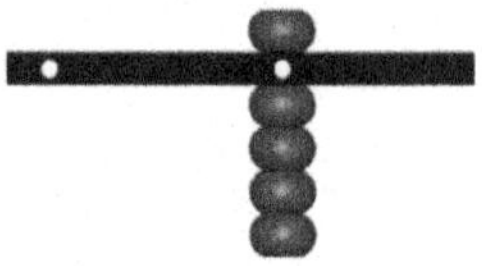	58 Number 4 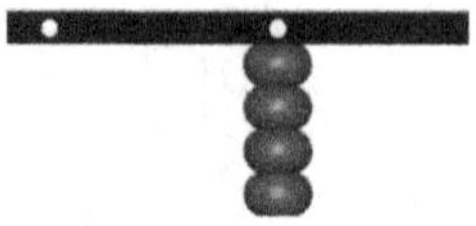	59 Number 5 	60 Number 11 

ANSWERS - 4

61 Number 6 	62 Number 9 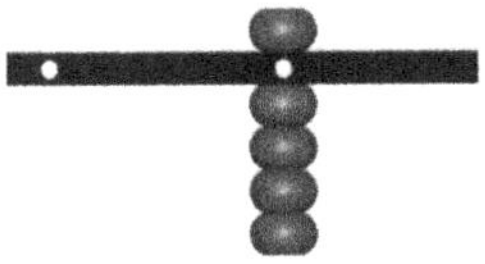	63 Number 9 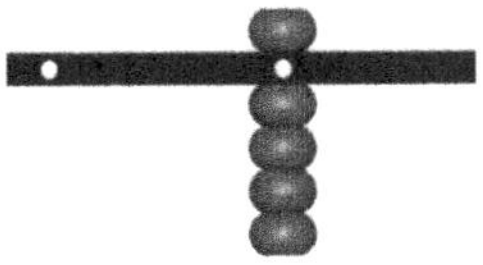	64 Number 11 
65 Number 13 	66 Number 10 	67 Number 14 	68 Number 11 
69 Number 13 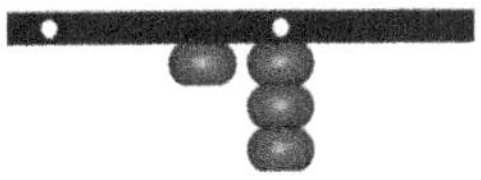	70 Number 8 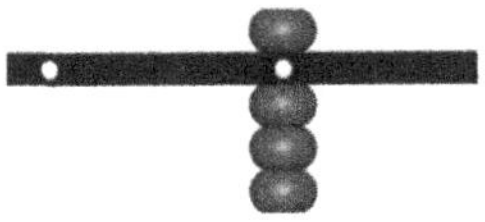	71 Number 12 	72 Number 12 
73 Number 16 	74 Number 17 	75 Number 13 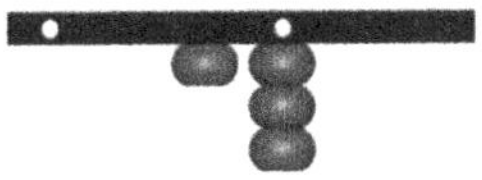	76 Number 15 
77 Number 18 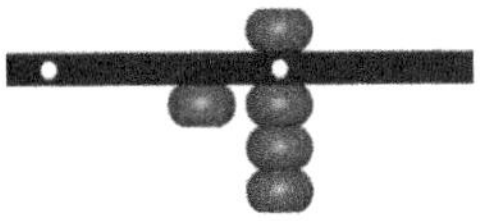	78 Number 15 	79 Number 17 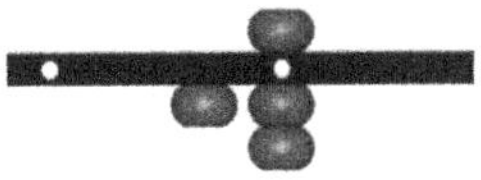	80 Number 20 

ANSWERS - 4

81 Number 25 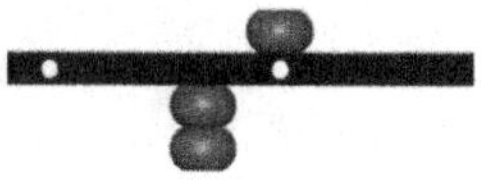	82 Number 36 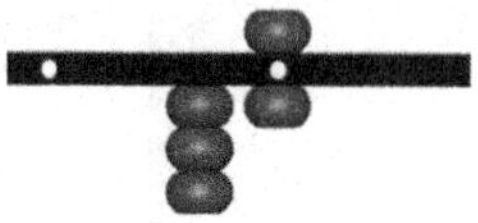	83 Number 24 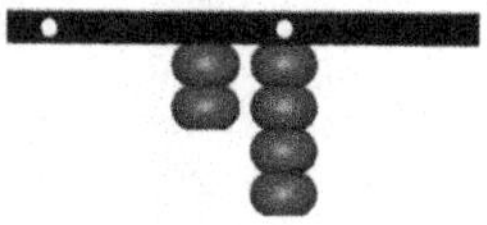	84 Number 17 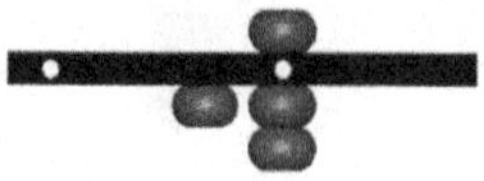
85 Number 21 	86 Number 48 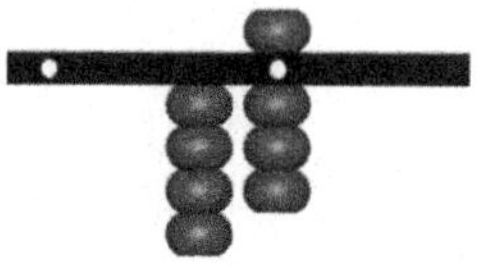	87 Number 27 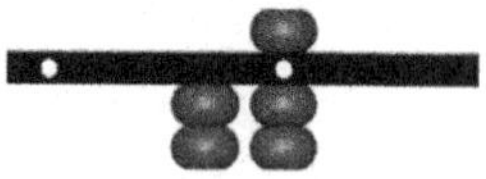	88 Number 33 
89 Number 36 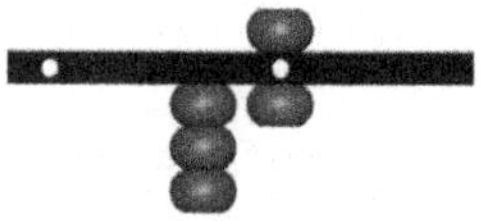	90 Number 33 	91 Number 60 	92 Number 52 
93 Number 52 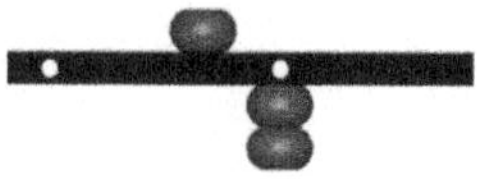	94 Number 61 	95 Number 66 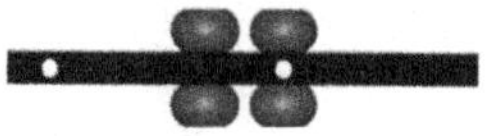	96 Number 81 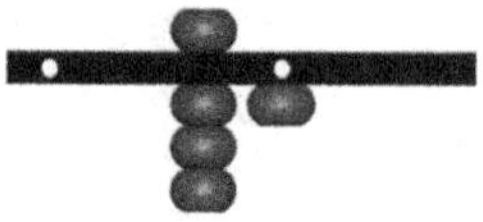
97 Number 77 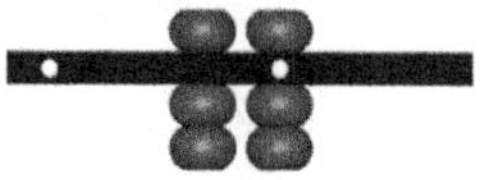	98 Number 83 	99 Number 99 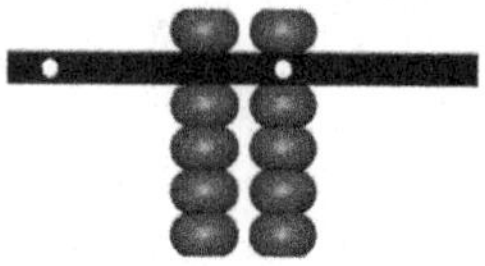	100 Number 78 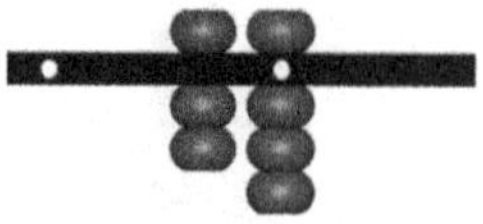

ANSWERS - 4

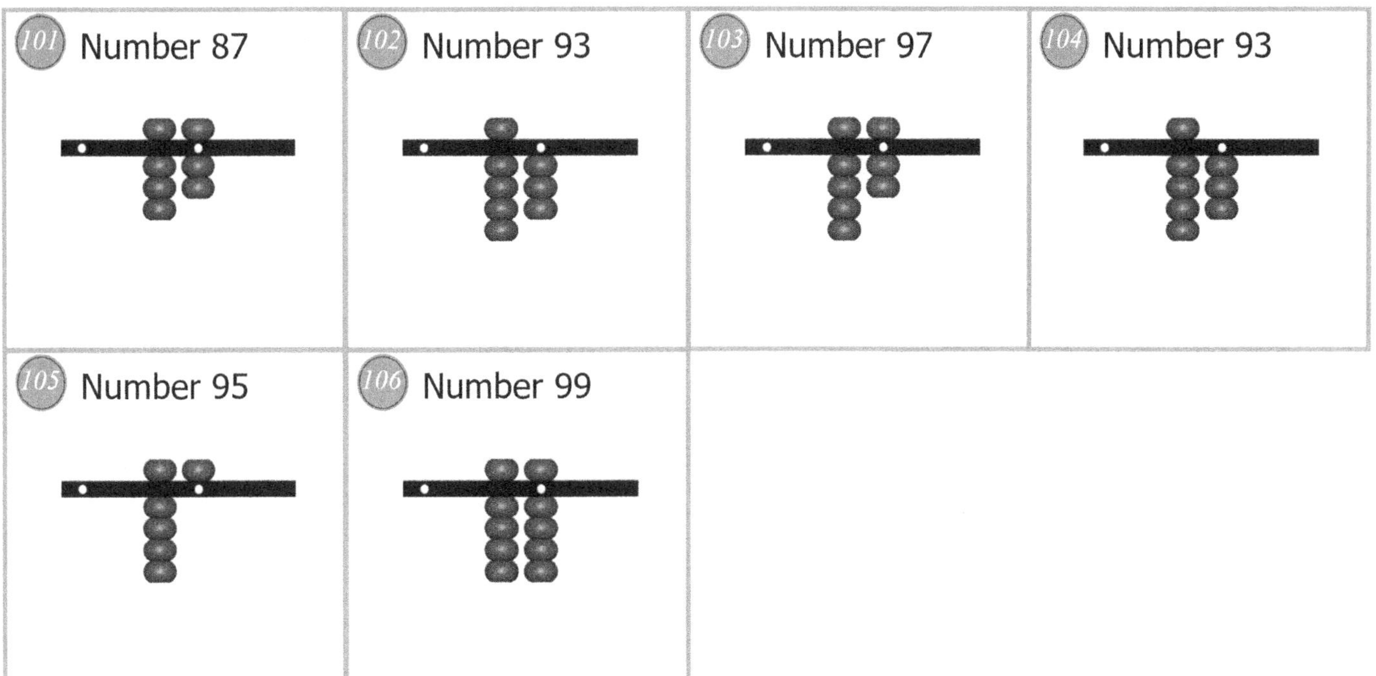

101
Number 87
102
Number 93
103
Number 97
104
Number 93
105
Number 95
106
Number 99

ANSWERS - 5

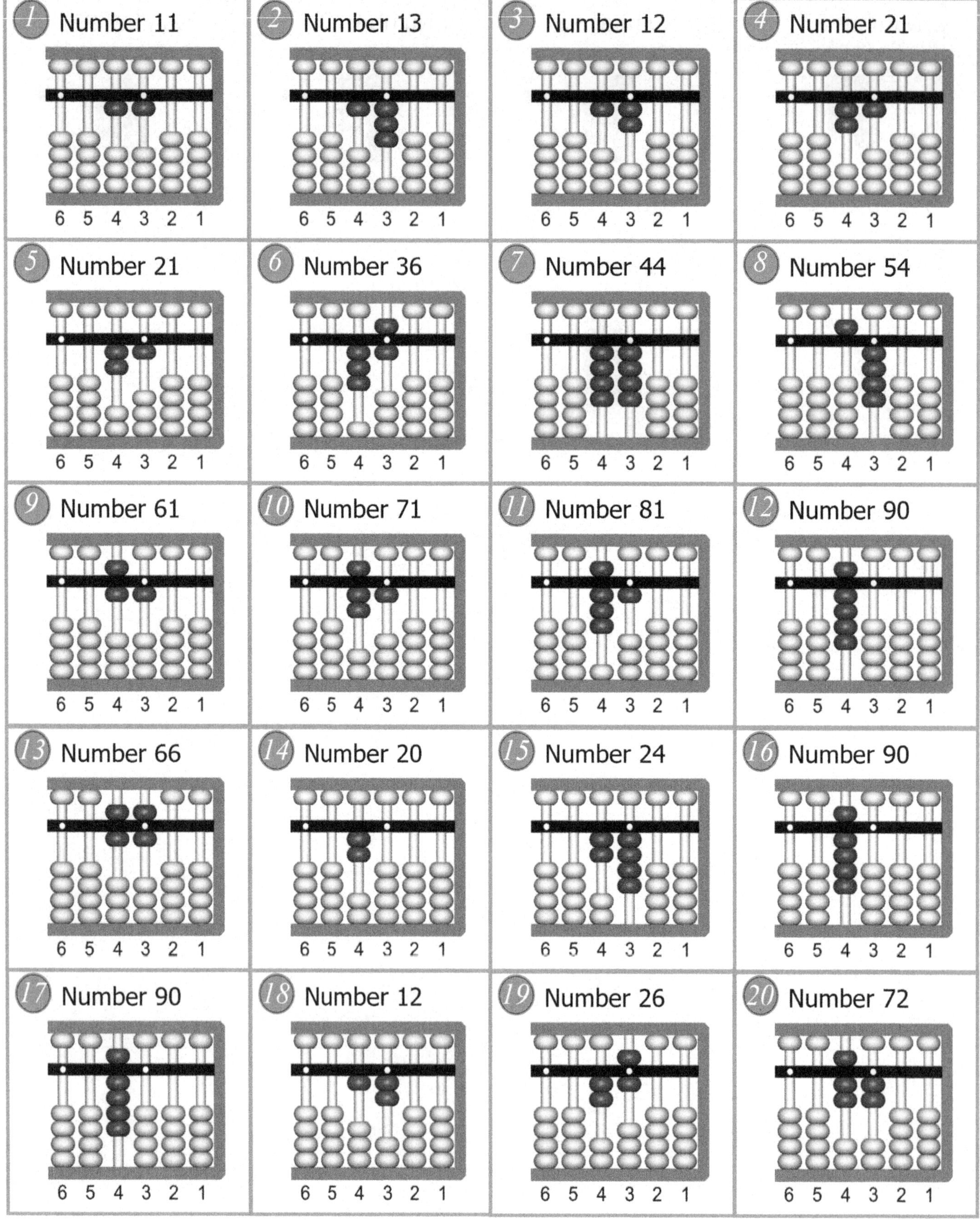
1 Number 11
6 5 4 3 2 1
2 Number 13
6 5 4 3 2 1
3 Number 12
6 5 4 3 2 1
4 Number 21
6 5 4 3 2 1
5 Number 21
6 5 4 3 2 1
6 Number 36
6 5 4 3 2 1
7 Number 44
6 5 4 3 2 1
8 Number 54
6 5 4 3 2 1
9 Number 61
6 5 4 3 2 1
10 Number 71
6 5 4 3 2 1
11 Number 81
6 5 4 3 2 1
12 Number 90
6 5 4 3 2 1
13 Number 66
6 5 4 3 2 1
14 Number 20
6 5 4 3 2 1
15 Number 24
6 5 4 3 2 1
16 Number 90
6 5 4 3 2 1
17 Number 90
6 5 4 3 2 1
18 Number 12
6 5 4 3 2 1
19 Number 26
6 5 4 3 2 1
20 Number 72
6 5 4 3 2 1

ANSWERS - 5

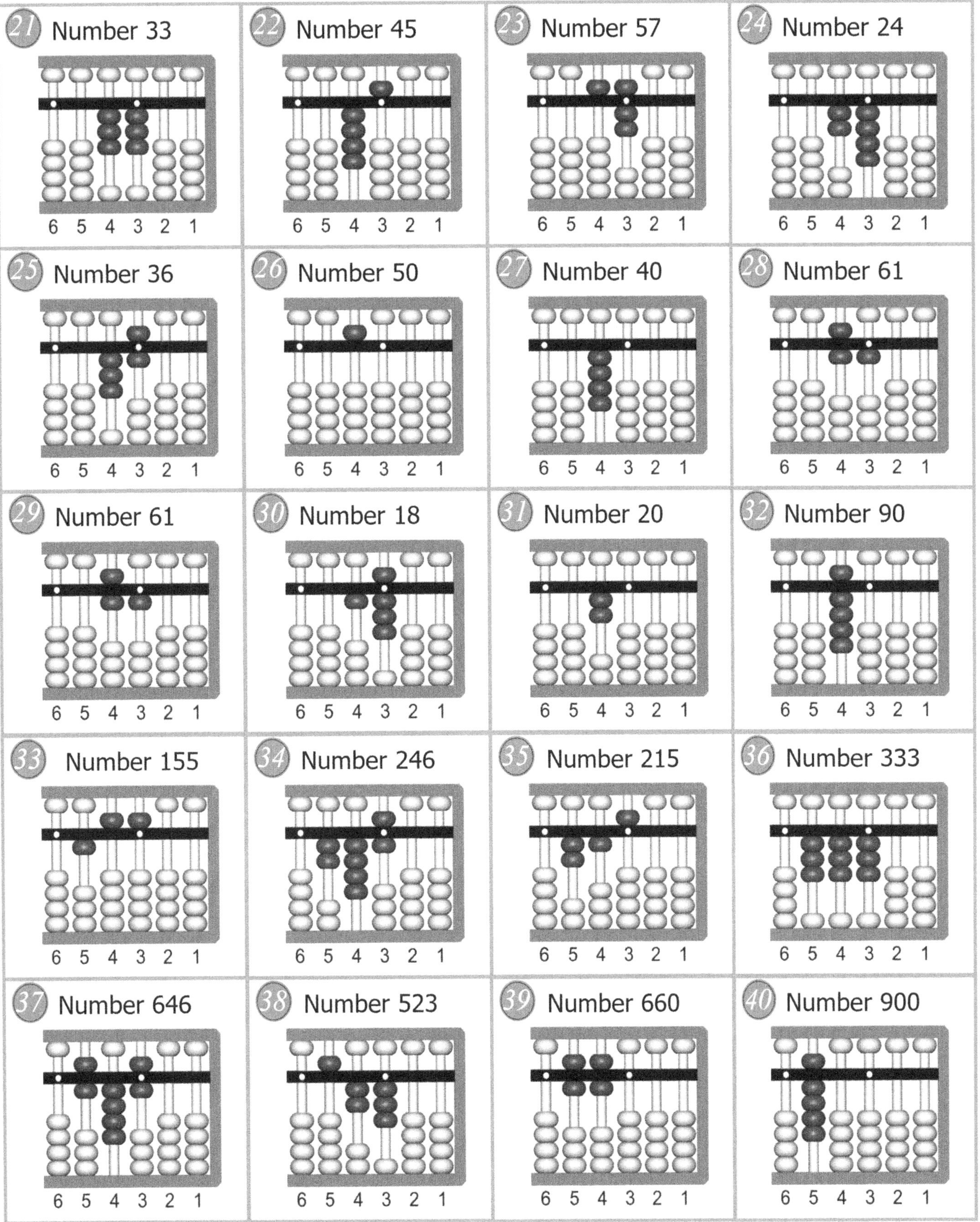

21 Number 33
6 5 4 3 2 1
22 Number 45
6 5 4 3 2 1
23 Number 57
6 5 4 3 2 1
24 Number 24
6 5 4 3 2 1
25 Number 36
6 5 4 3 2 1
26 Number 50
6 5 4 3 2 1
27 Number 40
6 5 4 3 2 1
28 Number 61
6 5 4 3 2 1
29 Number 61
6 5 4 3 2 1
30 Number 18
6 5 4 3 2 1
31 Number 20
6 5 4 3 2 1
32 Number 90
6 5 4 3 2 1
33 Number 155
6 5 4 3 2 1
34 Number 246
6 5 4 3 2 1
35 Number 215
6 5 4 3 2 1
36 Number 333
6 5 4 3 2 1
37 Number 646
6 5 4 3 2 1
38 Number 523
6 5 4 3 2 1
39 Number 660
6 5 4 3 2 1
40 Number 900
6 5 4 3 2 1

ANSWERS - 5

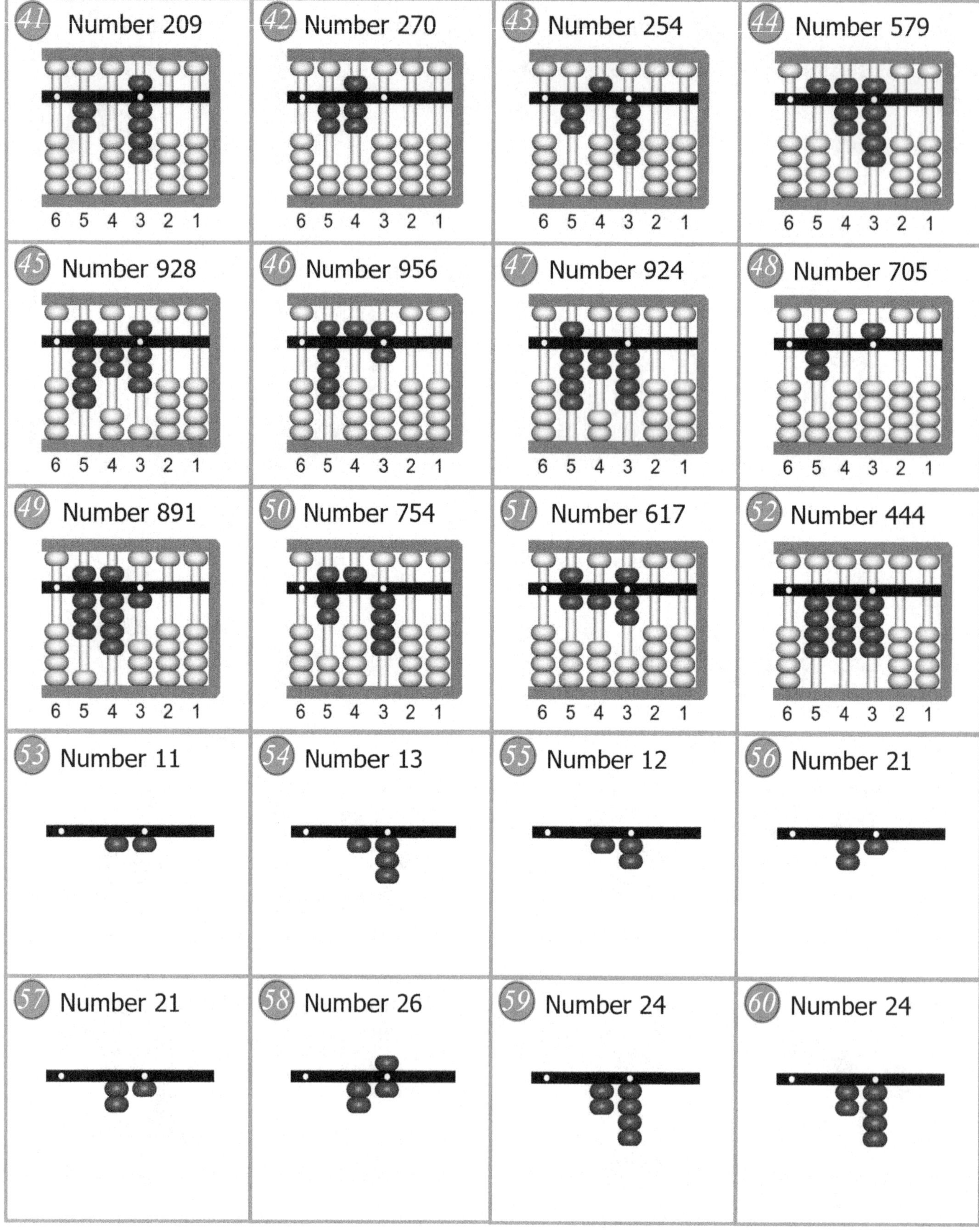
41 Number 209
6 5 4 3 2 1
42 Number 270
6 5 4 3 2 1
43 Number 254
6 5 4 3 2 1
44 Number 579
6 5 4 3 2 1
45 Number 928
6 5 4 3 2 1
46 Number 956
6 5 4 3 2 1
47 Number 924
6 5 4 3 2 1
48 Number 705
6 5 4 3 2 1
49 Number 891
6 5 4 3 2 1
50 Number 754
6 5 4 3 2 1
51 Number 617
6 5 4 3 2 1
52 Number 444
6 5 4 3 2 1
53 Number 11
54 Number 13
55 Number 12
56 Number 21
57 Number 21
58 Number 26
59 Number 24
60 Number 24

ANSWERS - 5

61 Number 31	62 Number 32	63 Number 23	64 Number 18
65 Number 27	66 Number 24	67 Number 34	68 Number 35
69 Number 43	70 Number 34	71 Number 31	72 Number 61
73 Number 64	74 Number 72	75 Number 77	76 Number 75
77 Number 81	78 Number 81	79 Number 87	80 Number 81

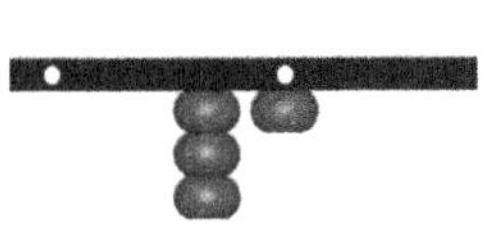

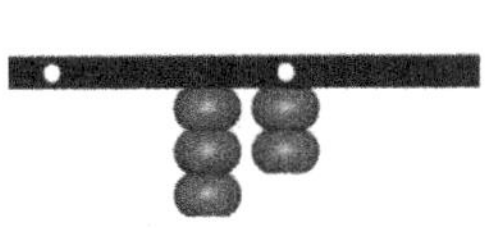

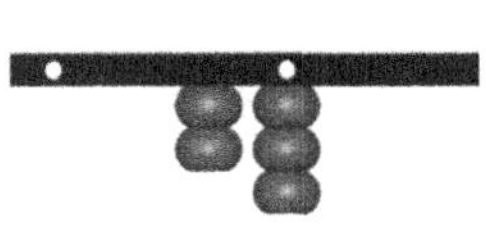

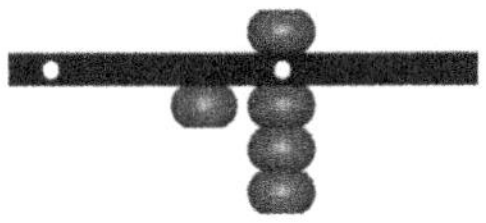

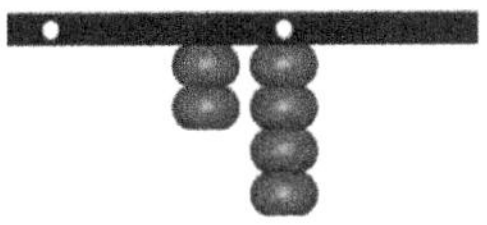

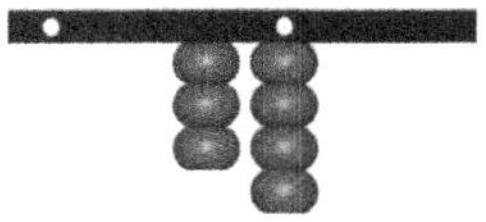

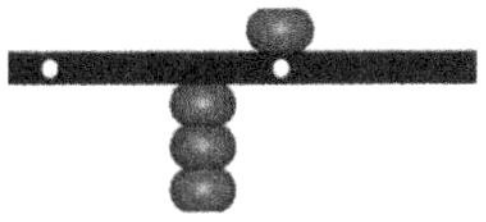

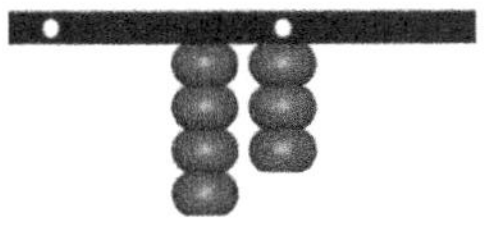

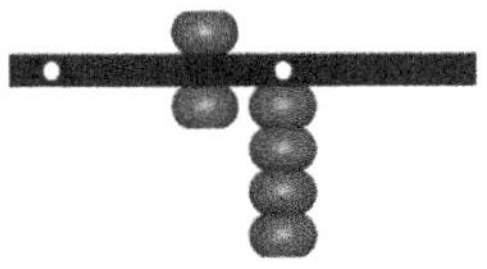

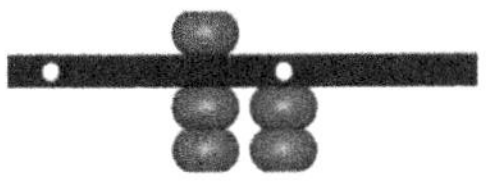

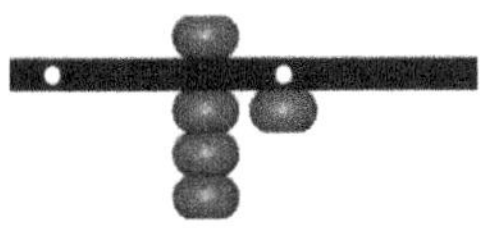

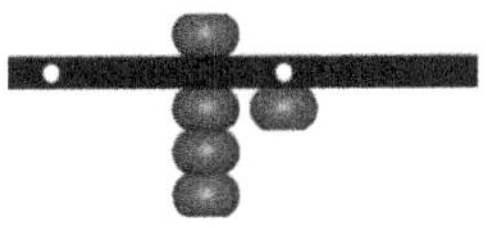

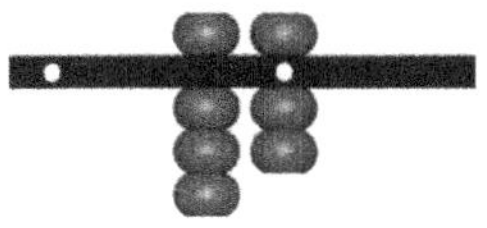

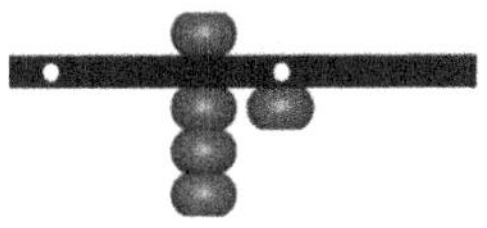

ANSWERS - 5

81 Number 61 	82 Number 43 	83 Number 23 	84 Number 41 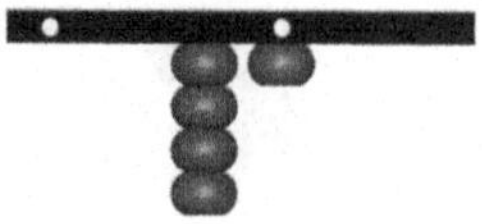
85 Number 22 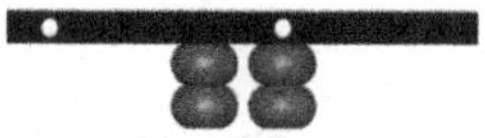	86 Number 20 	87 Number 82 	88 Number 91 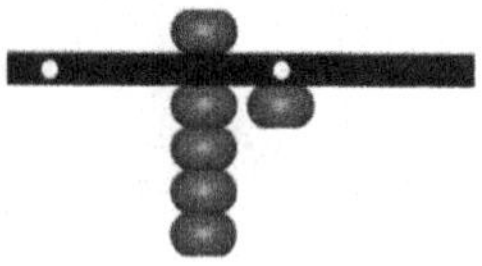
89 Number 74 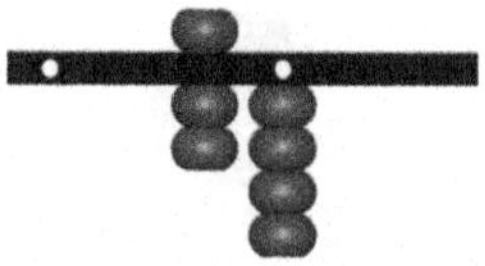	90 Number 44 	91 Number 21 	92 Number 62 
93 Number 74 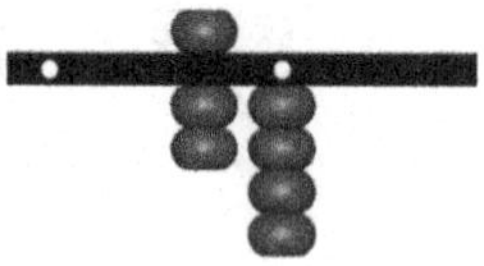	94 Number 80 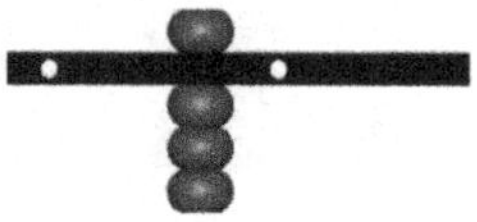	95 Number 91 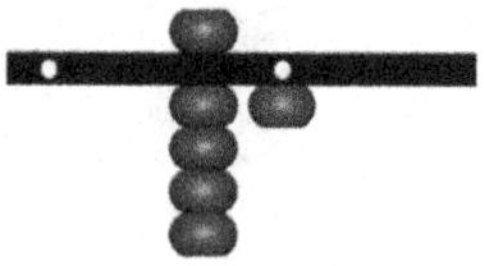	96 Number 99 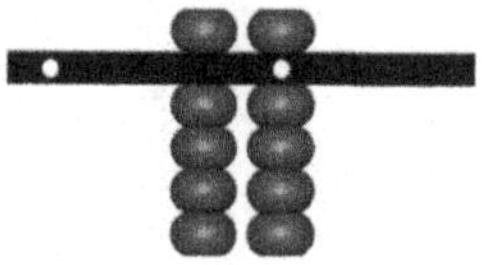
97 Number 60 	98 Number 72 	99 Number 86 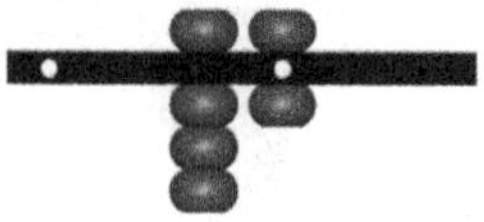	100 Number 93 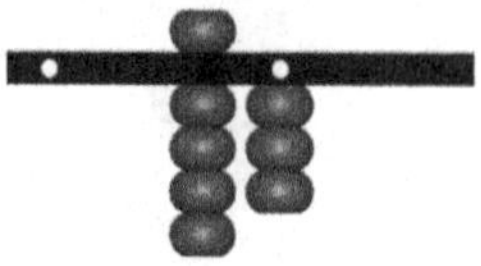

ANSWERS - 5

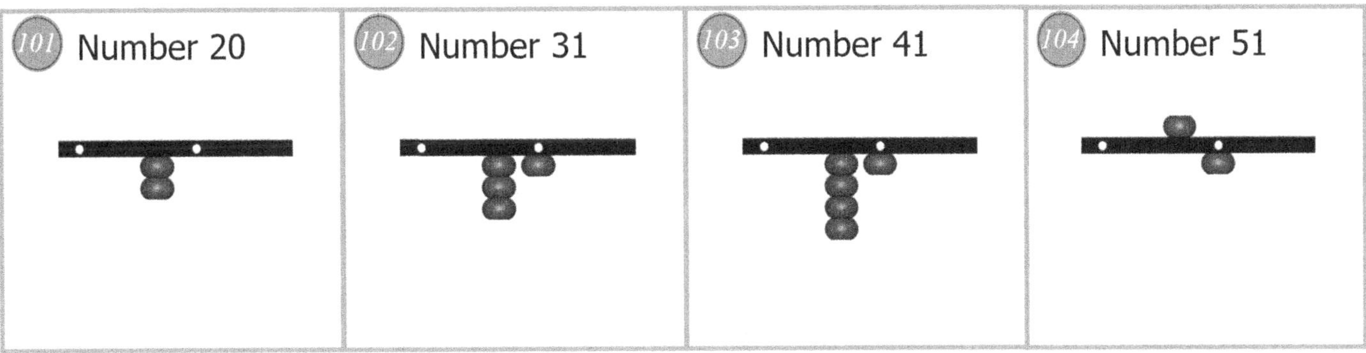

ANSWERS - 6

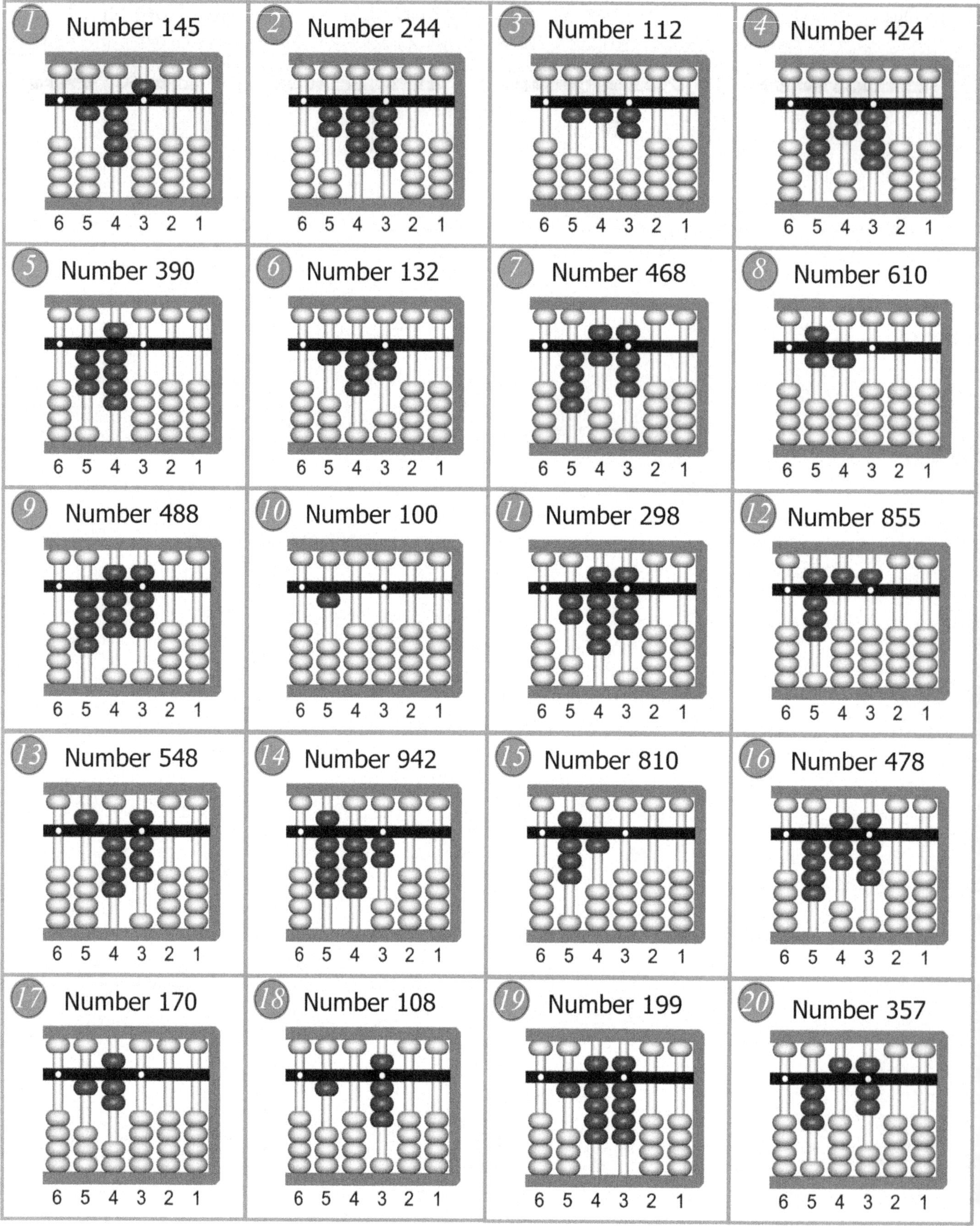
1 Number 145
6 5 4 3 2 1
2 Number 244
6 5 4 3 2 1
3 Number 112
6 5 4 3 2 1
4 Number 424
6 5 4 3 2 1
5 Number 390
6 5 4 3 2 1
6 Number 132
6 5 4 3 2 1
7 Number 468
6 5 4 3 2 1
8 Number 610
6 5 4 3 2 1
9 Number 488
6 5 4 3 2 1
10 Number 100
6 5 4 3 2 1
11 Number 298
6 5 4 3 2 1
12 Number 855
6 5 4 3 2 1
13 Number 548
6 5 4 3 2 1
14 Number 942
6 5 4 3 2 1
15 Number 810
6 5 4 3 2 1
16 Number 478
6 5 4 3 2 1
17 Number 170
6 5 4 3 2 1
18 Number 108
6 5 4 3 2 1
19 Number 199
6 5 4 3 2 1
20 Number 357
6 5 4 3 2 1

ANSWERS - 6

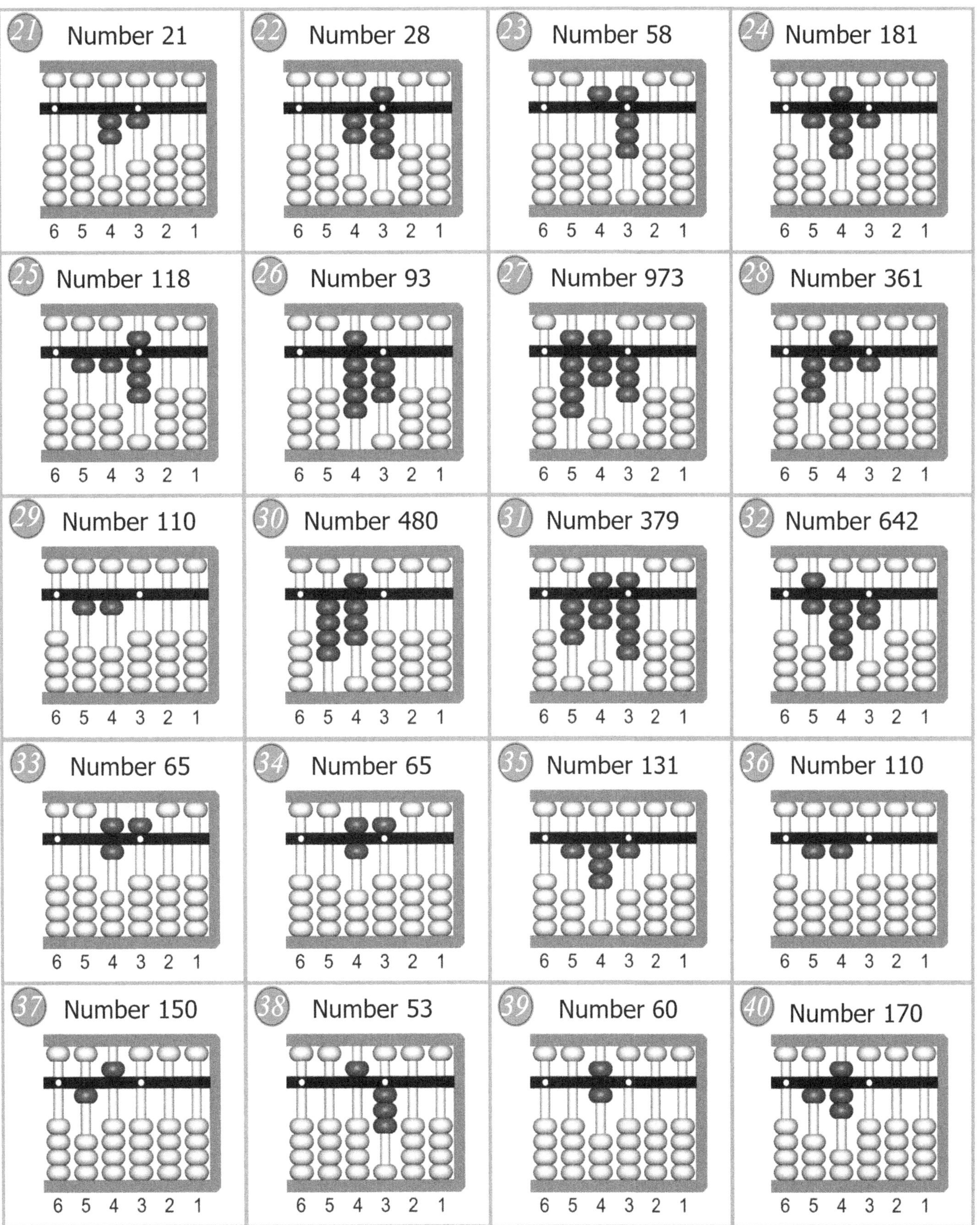
21
Number 21
6 5 4 3 2 1
22
Number 28
6 5 4 3 2 1
23
Number 58
6 5 4 3 2 1
24
Number 181
6 5 4 3 2 1
25
Number 118
6 5 4 3 2 1
26
Number 93
6 5 4 3 2 1
27
Number 973
6 5 4 3 2 1
28
Number 361
6 5 4 3 2 1
29
Number 110
6 5 4 3 2 1
30
Number 480
6 5 4 3 2 1
31
Number 379
6 5 4 3 2 1
32
Number 642
6 5 4 3 2 1
33
Number 65
6 5 4 3 2 1
34
Number 65
6 5 4 3 2 1
35
Number 131
6 5 4 3 2 1
36
Number 110
6 5 4 3 2 1
37
Number 150
6 5 4 3 2 1
38
Number 53
6 5 4 3 2 1
39
Number 60
6 5 4 3 2 1
40
Number 170
6 5 4 3 2 1

ANSWERS - 6

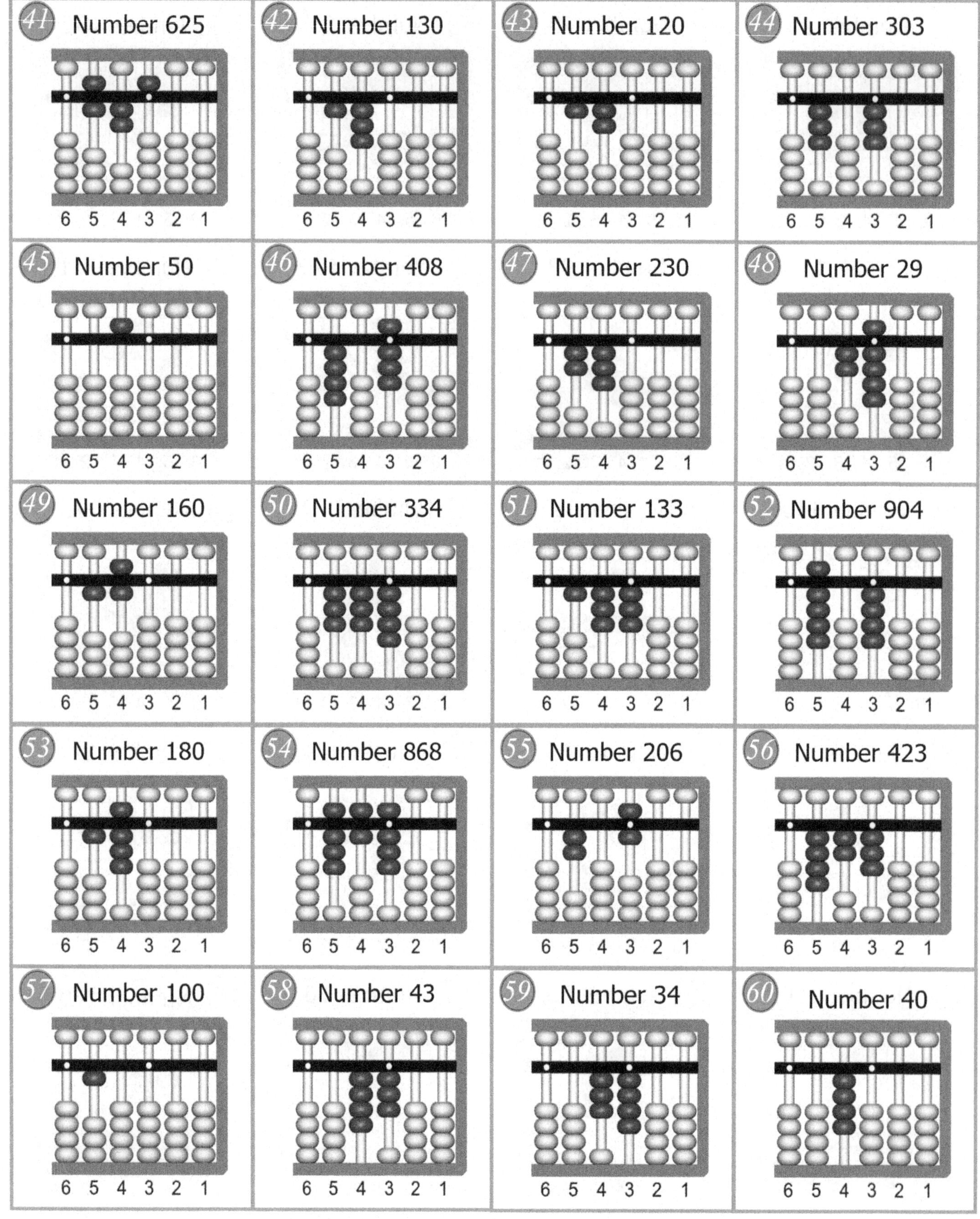
41 Number 625
6 5 4 3 2 1
42 Number 130
6 5 4 3 2 1
43 Number 120
6 5 4 3 2 1
44 Number 303
6 5 4 3 2 1
45 Number 50
6 5 4 3 2 1
46 Number 408
6 5 4 3 2 1
47 Number 230
6 5 4 3 2 1
48 Number 29
6 5 4 3 2 1
49 Number 160
6 5 4 3 2 1
50 Number 334
6 5 4 3 2 1
51 Number 133
6 5 4 3 2 1
52 Number 904
6 5 4 3 2 1
53 Number 180
6 5 4 3 2 1
54 Number 868
6 5 4 3 2 1
55 Number 206
6 5 4 3 2 1
56 Number 423
6 5 4 3 2 1
57 Number 100
6 5 4 3 2 1
58 Number 43
6 5 4 3 2 1
59 Number 34
6 5 4 3 2 1
60 Number 40
6 5 4 3 2 1

ANSWERS - 6

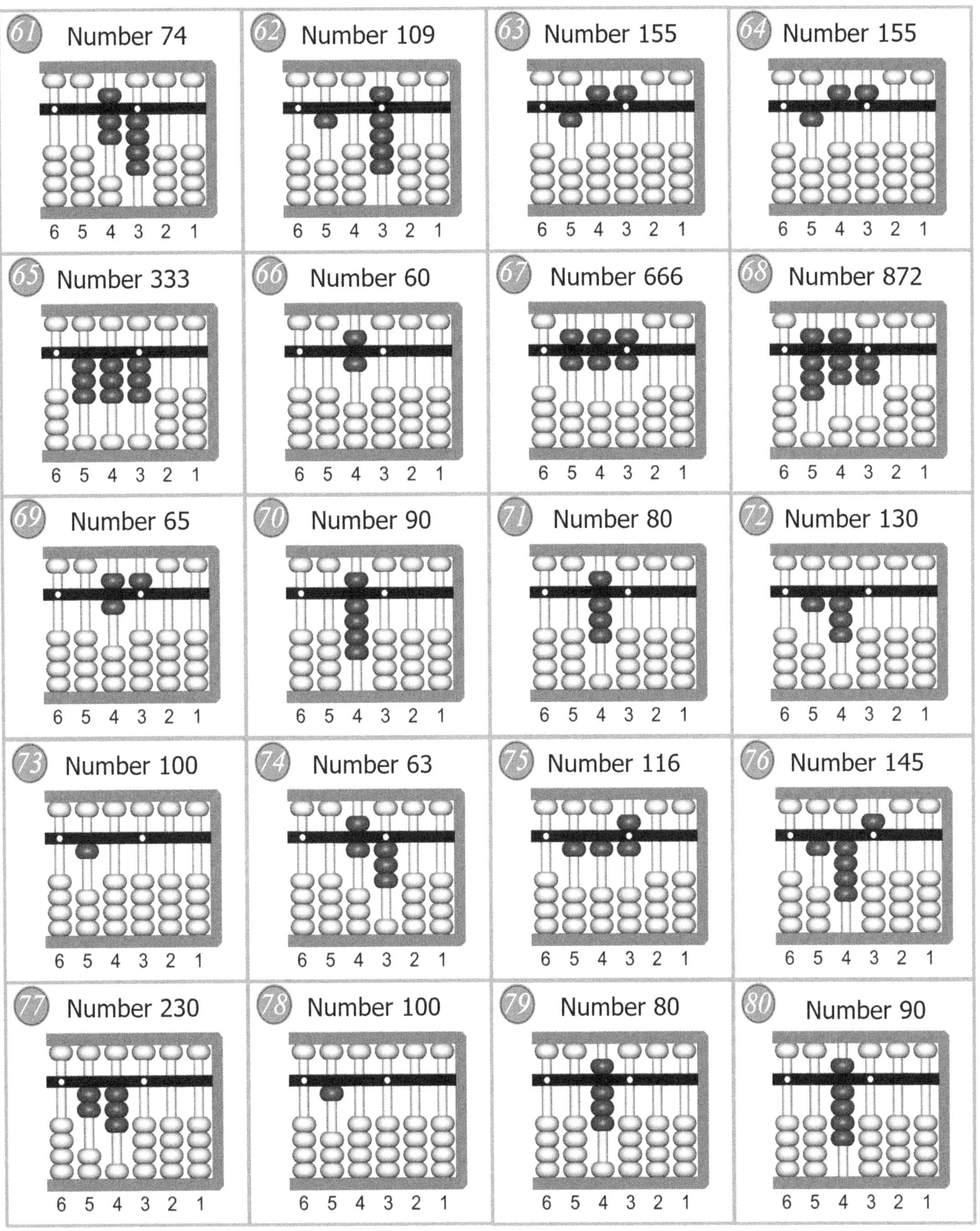
61
Number 74
6 5 4 3 2 1
62
Number 109
6 5 4 3 2 1
63
Number 155
6 5 4 3 2 1
64
Number 155
6 5 4 3 2 1
65
Number 333
6 5 4 3 2 1
66
Number 60
6 5 4 3 2 1
67
Number 666
6 5 4 3 2 1
68
Number 872
6 5 4 3 2 1
69
Number 65
6 5 4 3 2 1
70
Number 90
6 5 4 3 2 1
71
Number 80
6 5 4 3 2 1
72
Number 130
6 5 4 3 2 1
73
Number 100
6 5 4 3 2 1
74
Number 63
6 5 4 3 2 1
75
Number 116
6 5 4 3 2 1
76
Number 145
6 5 4 3 2 1
77
Number 230
6 5 4 3 2 1
78
Number 100
6 5 4 3 2 1
79
Number 80
6 5 4 3 2 1
80
Number 90
6 5 4 3 2 1

ANSWERS - 6

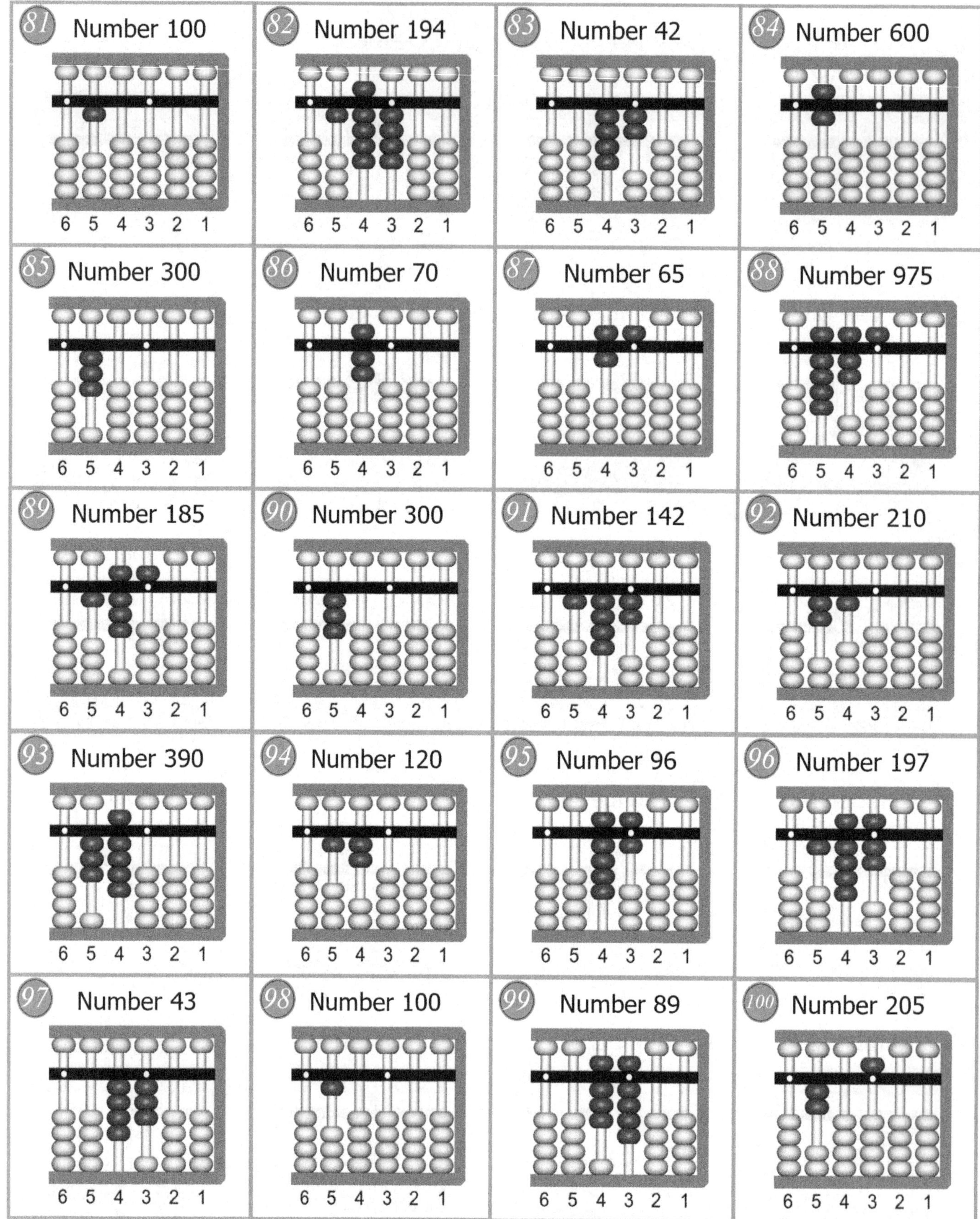
81 Number 100
6 5 4 3 2 1
82 Number 194
6 5 4 3 2 1
83 Number 42
6 5 4 3 2 1
84 Number 600
6 5 4 3 2 1
85 Number 300
6 5 4 3 2 1
86 Number 70
6 5 4 3 2 1
87 Number 65
6 5 4 3 2 1
88 Number 975
6 5 4 3 2 1
89 Number 185
6 5 4 3 2 1
90 Number 300
6 5 4 3 2 1
91 Number 142
6 5 4 3 2 1
92 Number 210
6 5 4 3 2 1
93 Number 390
6 5 4 3 2 1
94 Number 120
6 5 4 3 2 1
95 Number 96
6 5 4 3 2 1
96 Number 197
6 5 4 3 2 1
97 Number 43
6 5 4 3 2 1
98 Number 100
6 5 4 3 2 1
99 Number 89
6 5 4 3 2 1
100 Number 205
6 5 4 3 2 1

ANSWERS - 7

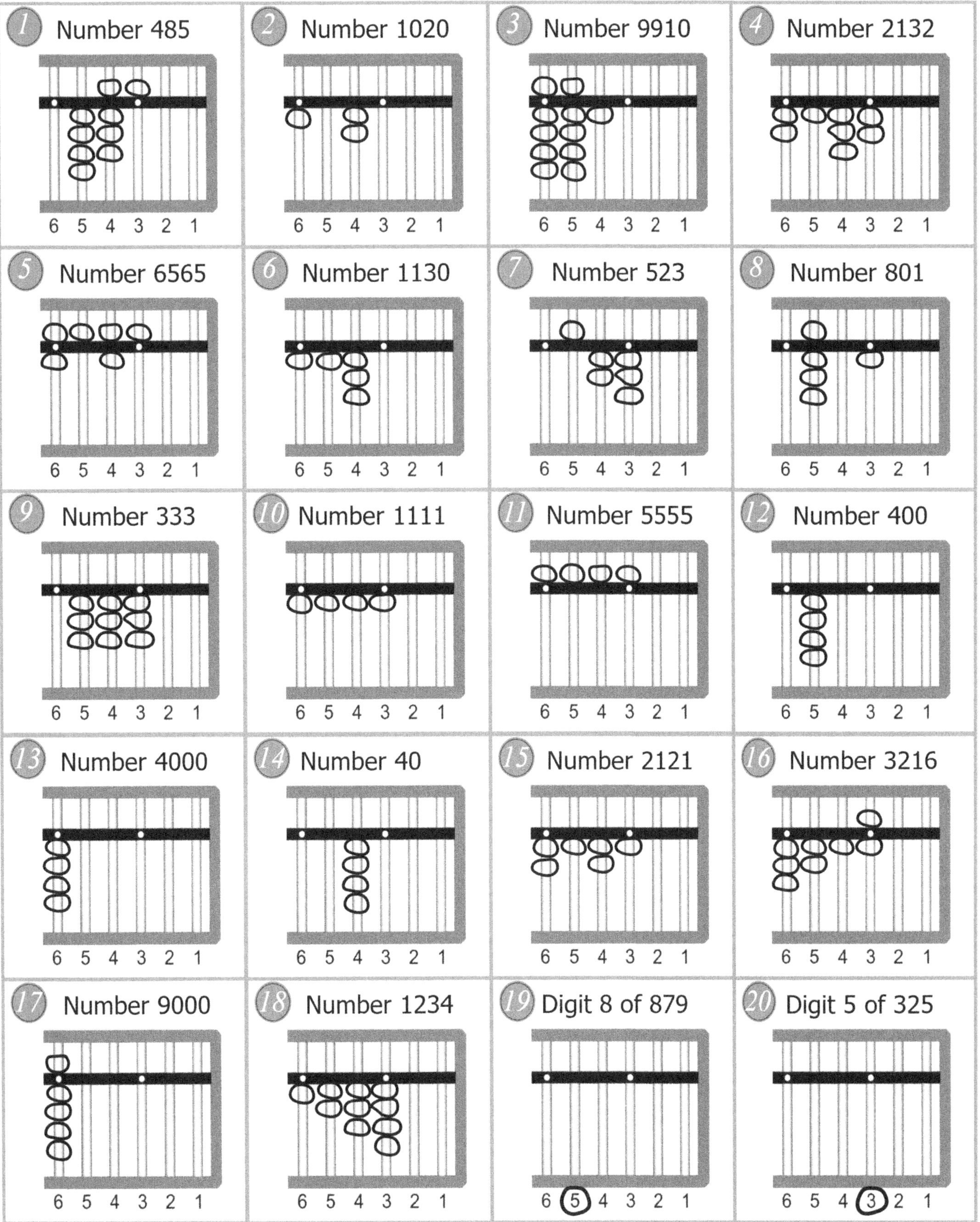
1 Number 485
2 Number 1020
3 Number 9910
4 Number 2132
6 5 4 3 2 1
5 Number 6565
6 Number 1130
7 Number 523
8 Number 801
6 5 4 3 2 1
9 Number 333
10 Number 1111
11 Number 5555
12 Number 400
6 5 4 3 2 1
13 Number 4000
14 Number 40
15 Number 2121
16 Number 3216
6 5 4 3 2 1
17 Number 9000
18 Number 1234
19 Digit 8 of 879
20 Digit 5 of 325
6 5 4 3 2 1

ANSWERS - 7

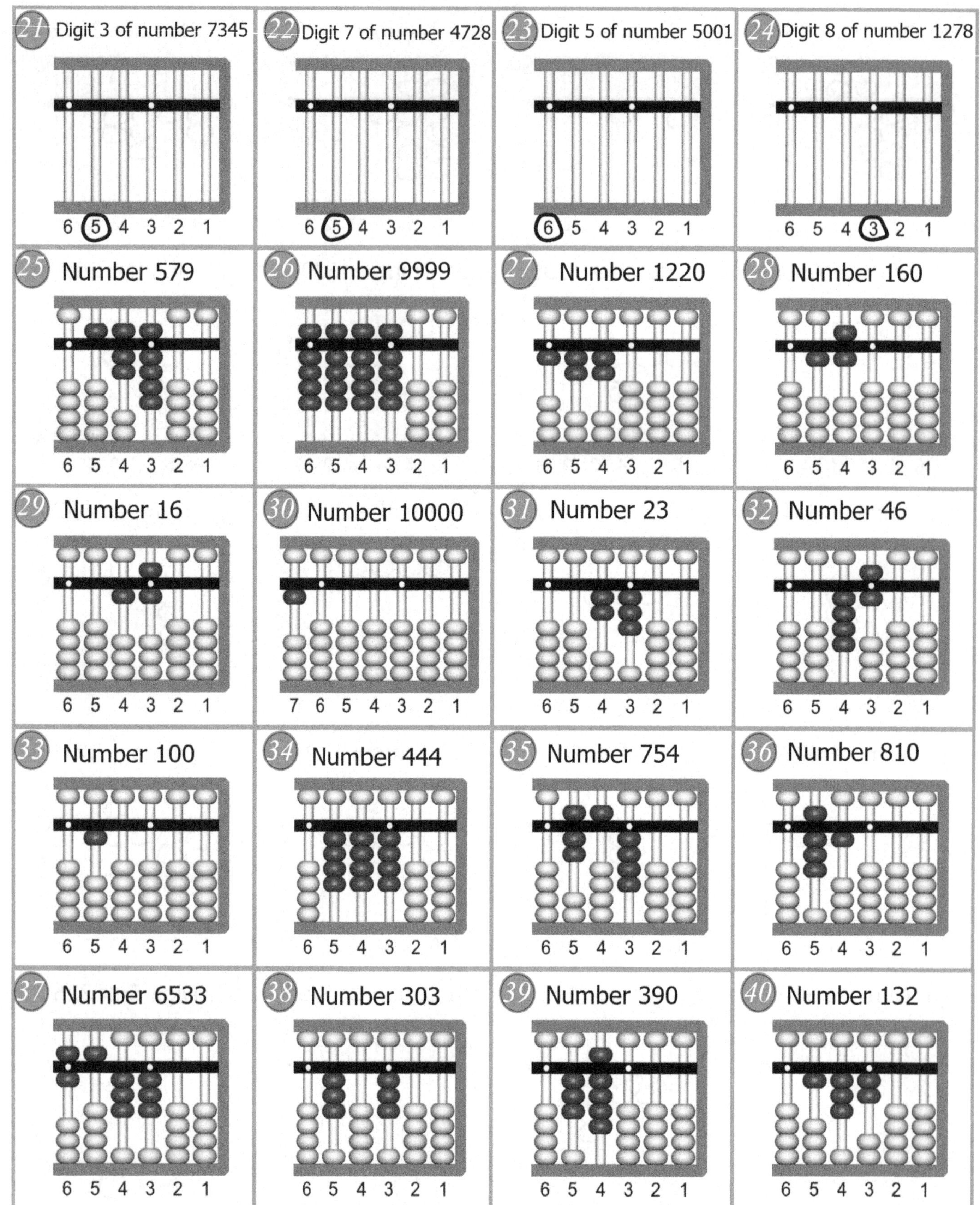

21 Digit 3 of number 7345
6 5 4 3 2 1
22 Digit 7 of number 4728
6 5 4 3 2 1
23 Digit 5 of number 5001
6 5 4 3 2 1
24 Digit 8 of number 1278
6 5 4 3 2 1
25 Number 579
6 5 4 3 2 1
26 Number 9999
6 5 4 3 2 1
27 Number 1220
6 5 4 3 2 1
28 Number 160
6 5 4 3 2 1
29 Number 16
6 5 4 3 2 1
30 Number 10000
7 6 5 4 3 2 1
31 Number 23
6 5 4 3 2 1
32 Number 46
6 5 4 3 2 1
33 Number 100
6 5 4 3 2 1
34 Number 444
6 5 4 3 2 1
35 Number 754
6 5 4 3 2 1
36 Number 810
6 5 4 3 2 1
37 Number 6533
6 5 4 3 2 1
38 Number 303
6 5 4 3 2 1
39 Number 390
6 5 4 3 2 1
40 Number 132
6 5 4 3 2 1

ANSWERS - 7

41 Number 6533	42 Number 244	43 Number 4	44 Number 50
45 Number 112	46 Number 33	47 Number 55	48 Number 55
49 Number 131	50 Number 133	51 Number 424	52 Number 13
53 Number 46	54 Number 33	55 Number 21	56 Number 23
57 Number 50	58 Number 160	59 Number 244	60 Number 334

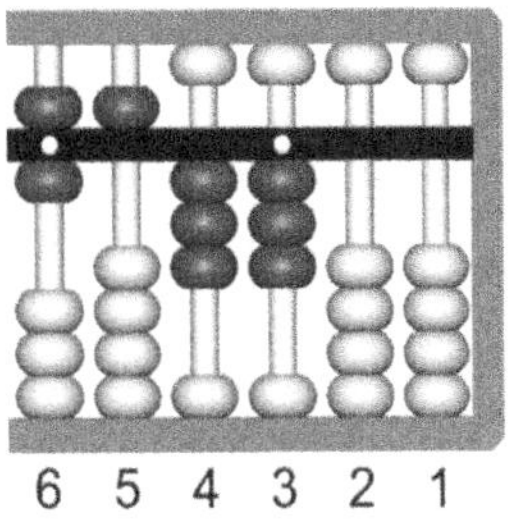

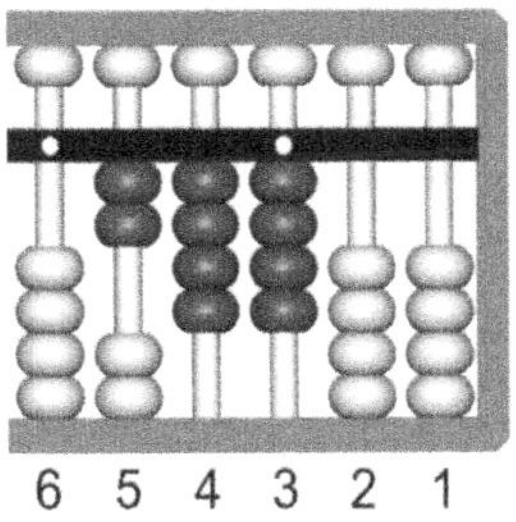

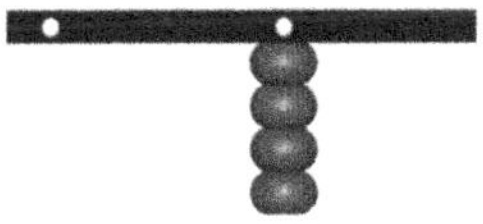

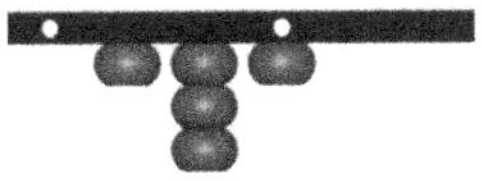

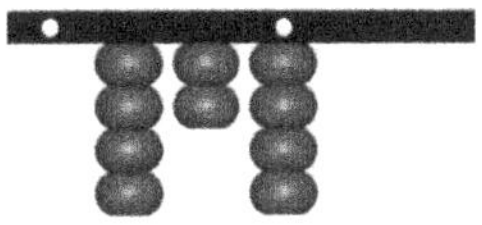

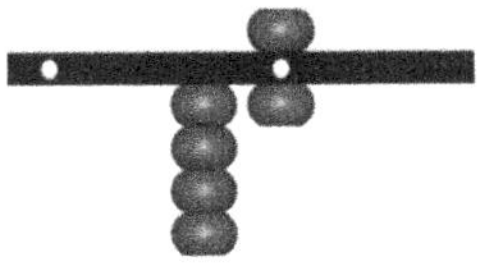

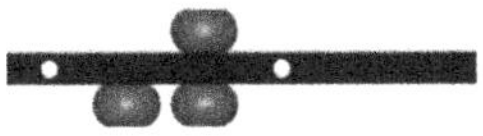

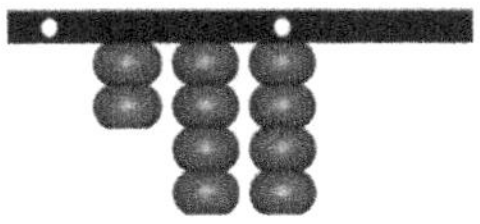

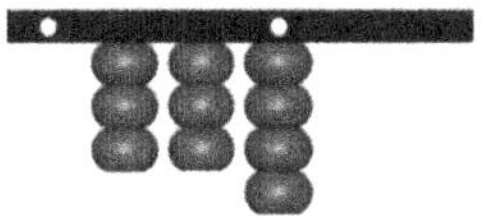

6 5 4 3 2 1

6 5 4 3 2 1

ANSWERS - 7

61 Number 523	62 Number 100	63 Number 10	64 Number 70
65 Number 254	66 Number 705	67 Number 194	68 Number 523
69 Number 868	70 Number 300	71 Number 104	72 Number 22
73 Number 210	74 Number 130		

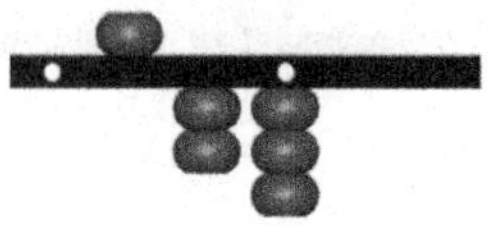

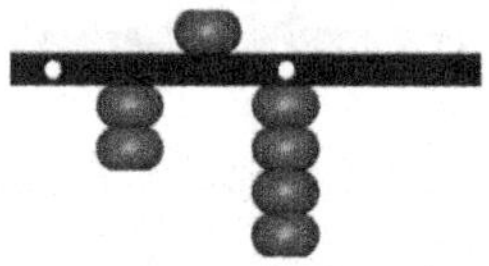

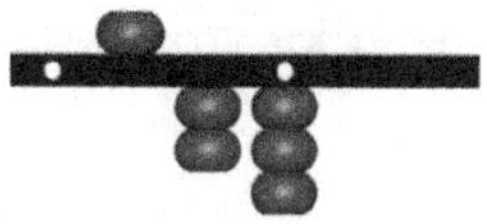

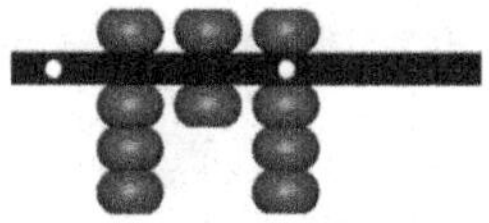

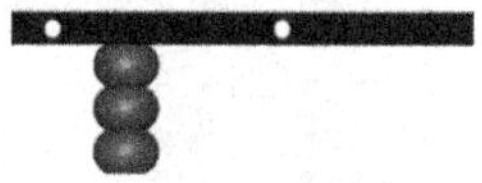

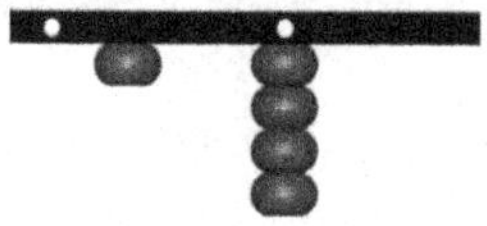

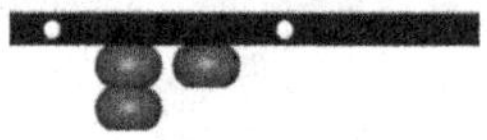

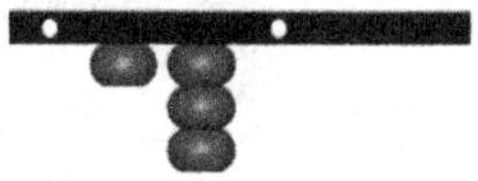

ANSWERS - 8

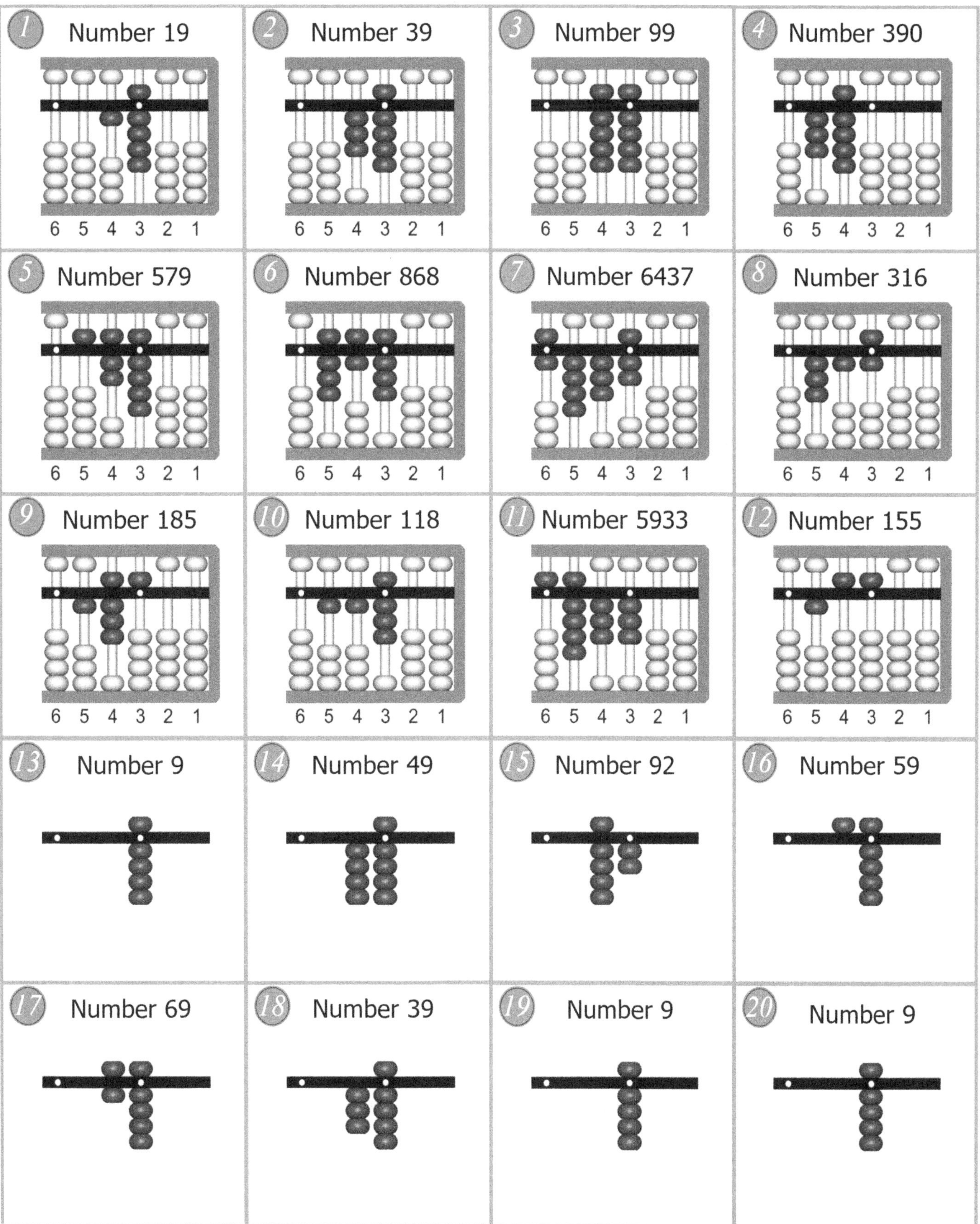

1 Number 19
6 5 4 3 2 1
2 Number 39
6 5 4 3 2 1
3 Number 99
6 5 4 3 2 1
4 Number 390
6 5 4 3 2 1
5 Number 579
6 5 4 3 2 1
6 Number 868
6 5 4 3 2 1
7 Number 6437
6 5 4 3 2 1
8 Number 316
6 5 4 3 2 1
9 Number 185
6 5 4 3 2 1
10 Number 118
6 5 4 3 2 1
11 Number 5933
6 5 4 3 2 1
12 Number 155
6 5 4 3 2 1
13 Number 9
14 Number 49
15 Number 92
16 Number 59
17 Number 69
18 Number 39
19 Number 9
20 Number 9

ANSWERS - 8

21 Number 9	22 Number 9	23 Number 4	24 Number 84
25 Number 39	26 Number 69	27 Number 59	28 Number 99
29 Number 39	30 Number 59	31 Number 49	32 Number 104
33 Number 90	34 Number 79	35 Number 19	36 Number 18
37 Number 9	38 Number 80	39 Number 29	40 Number 90

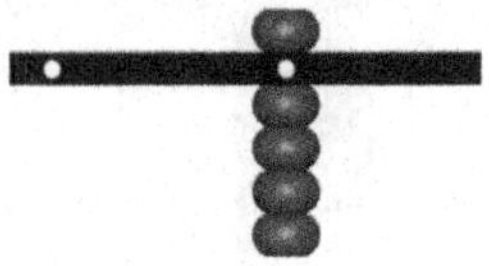

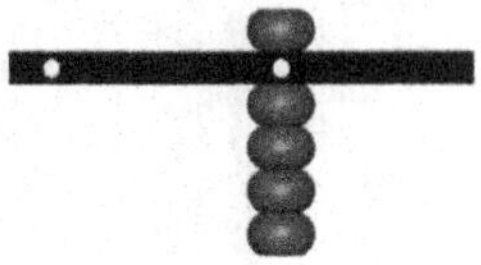

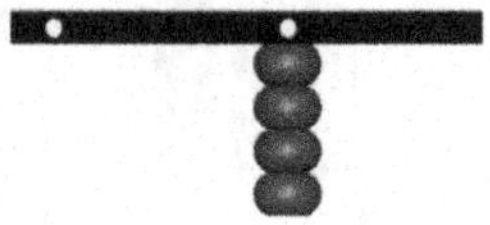

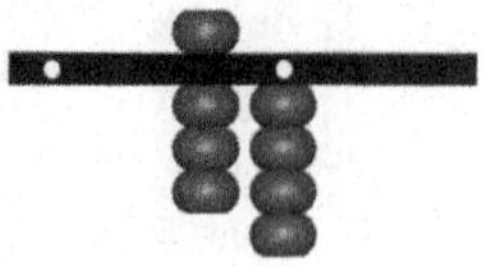

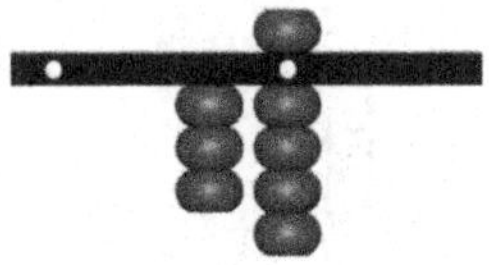

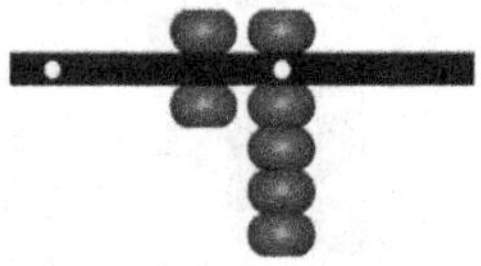

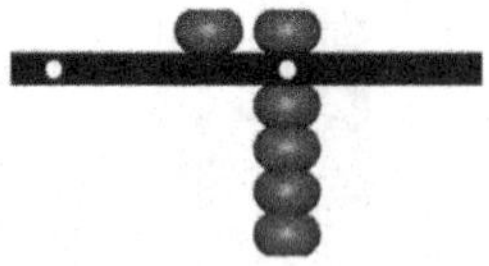

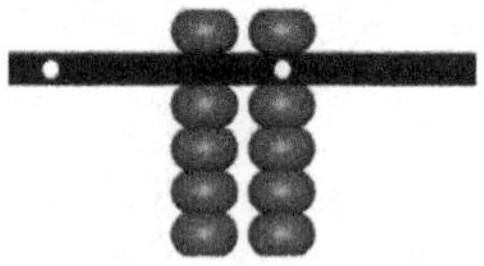

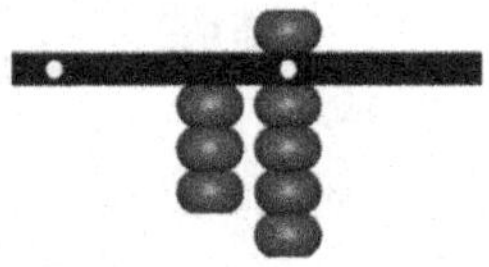

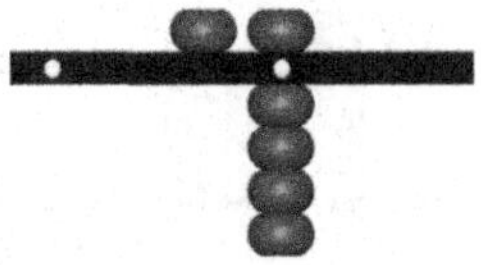

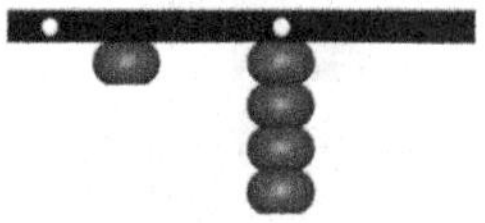

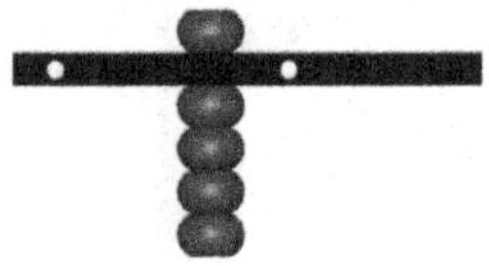

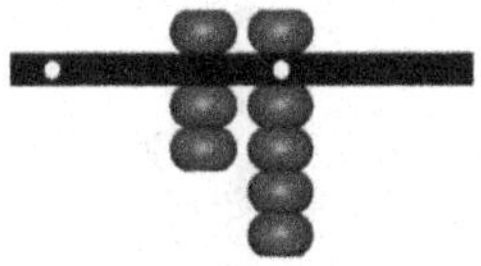

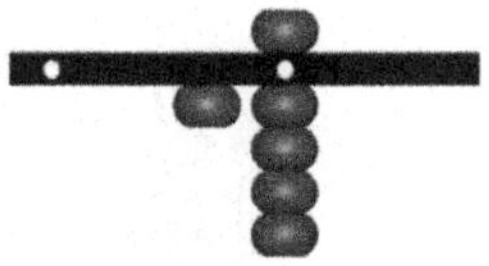

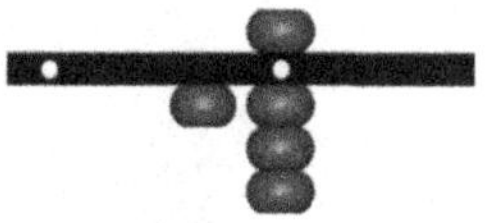

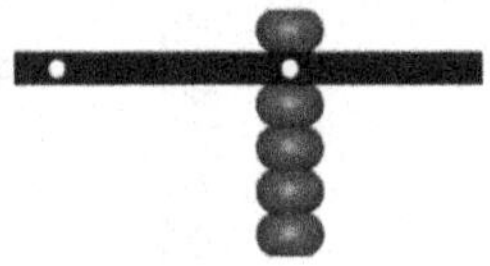

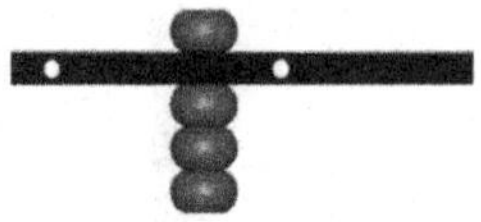

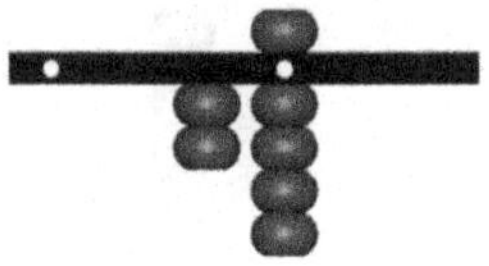

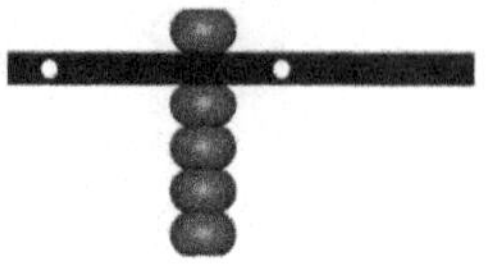

ANSWERS - 9

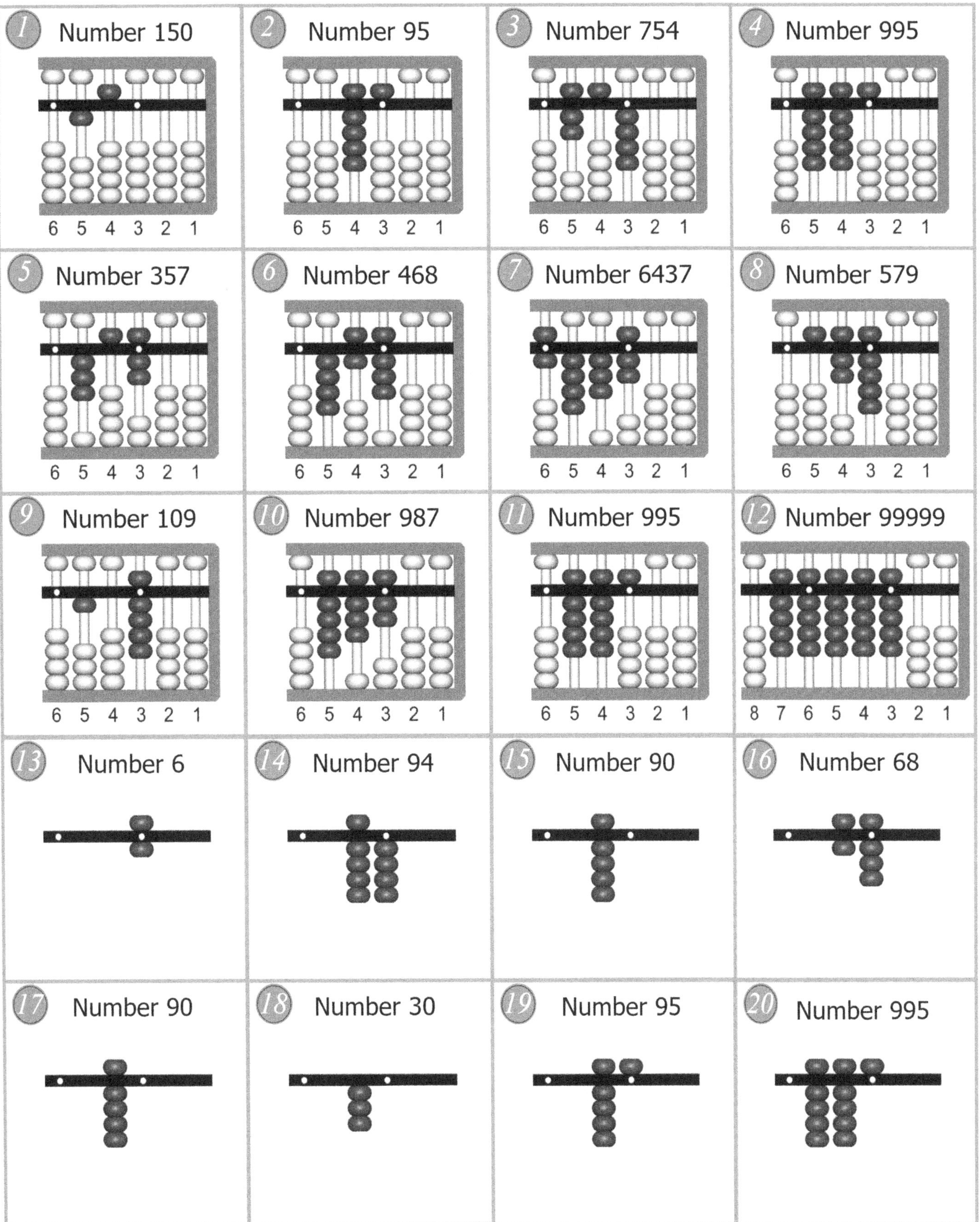
1 Number 150
6 5 4 3 2 1
2 Number 95
6 5 4 3 2 1
3 Number 754
6 5 4 3 2 1
4 Number 995
6 5 4 3 2 1
5 Number 357
6 5 4 3 2 1
6 Number 468
6 5 4 3 2 1
7 Number 6437
6 5 4 3 2 1
8 Number 579
6 5 4 3 2 1
9 Number 109
6 5 4 3 2 1
10 Number 987
6 5 4 3 2 1
11 Number 995
6 5 4 3 2 1
12 Number 99999
8 7 6 5 4 3 2 1
13 Number 6
14 Number 94
15 Number 90
16 Number 68
17 Number 90
18 Number 30
19 Number 95
20 Number 995

ANSWERS - 9

21 Number 204 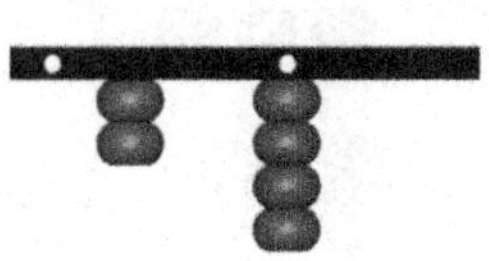	22 Number 9999 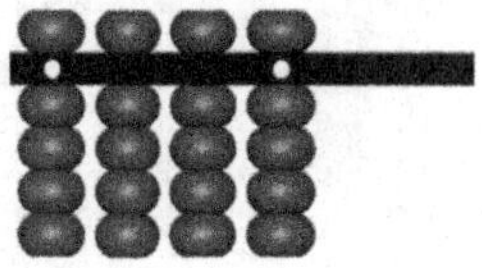	23 Number 120 	24 Number 72 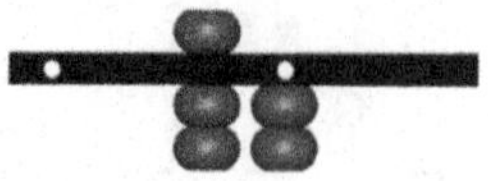
25 Number 40 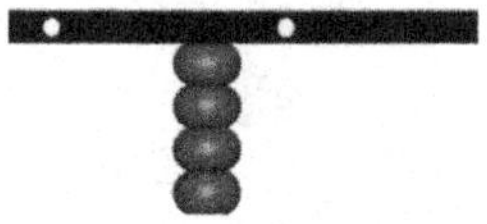	26 Number 72 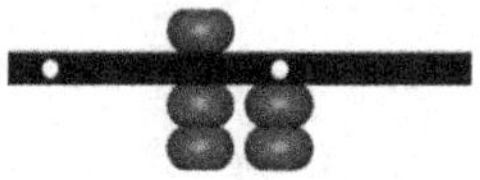	27 Number 52 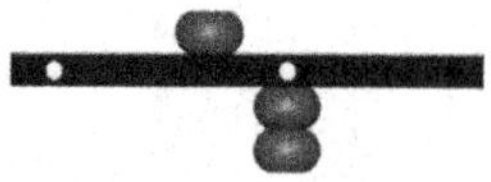	28 Number 109 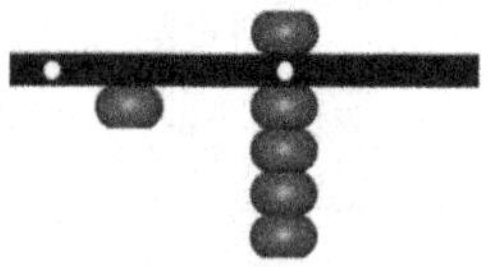
29 Number 119 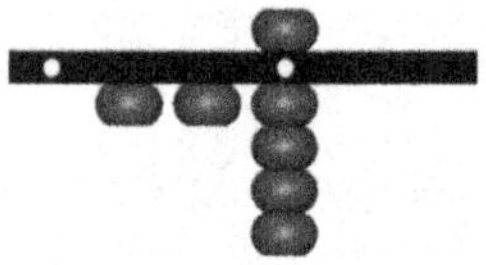	30 Number 91 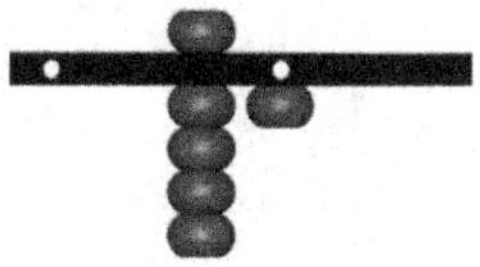	31 Number 191 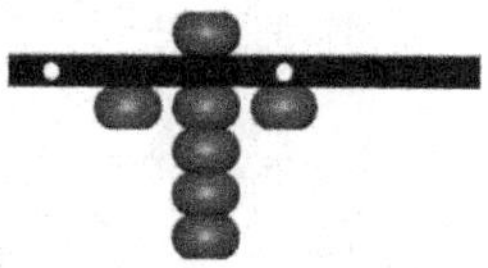	32 Number 390 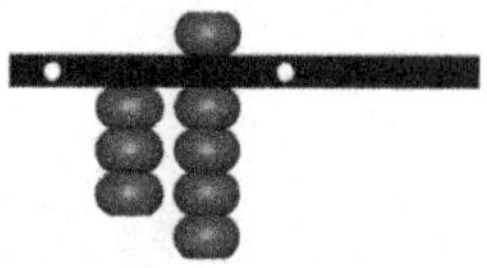
33 Number 548 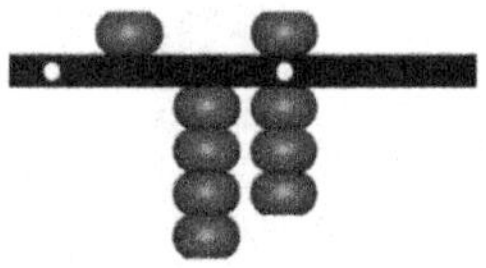	34 Number 63 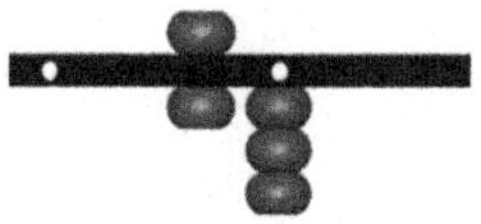	35 Number 40 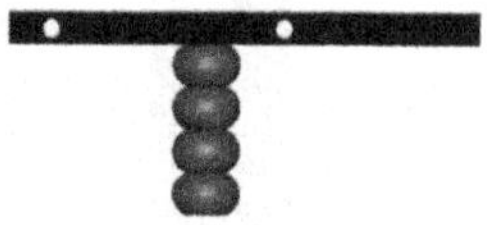	36 Number 334 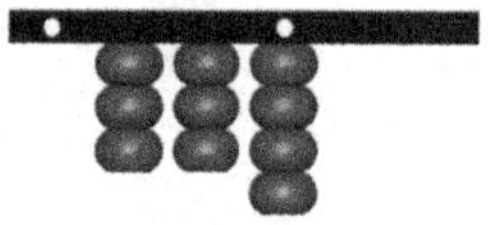
37 Number 185 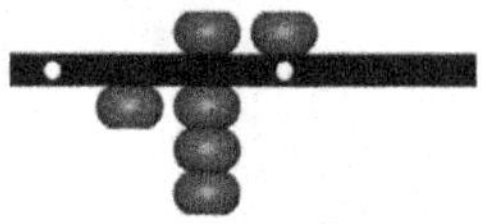	38 Number 78 	39 Number 377 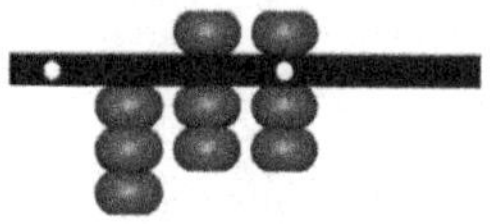	40 Number 213 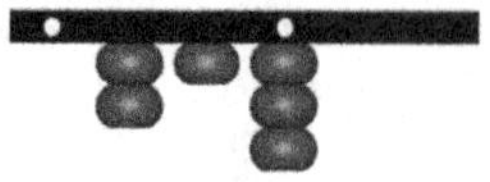

ANSWERS - 10

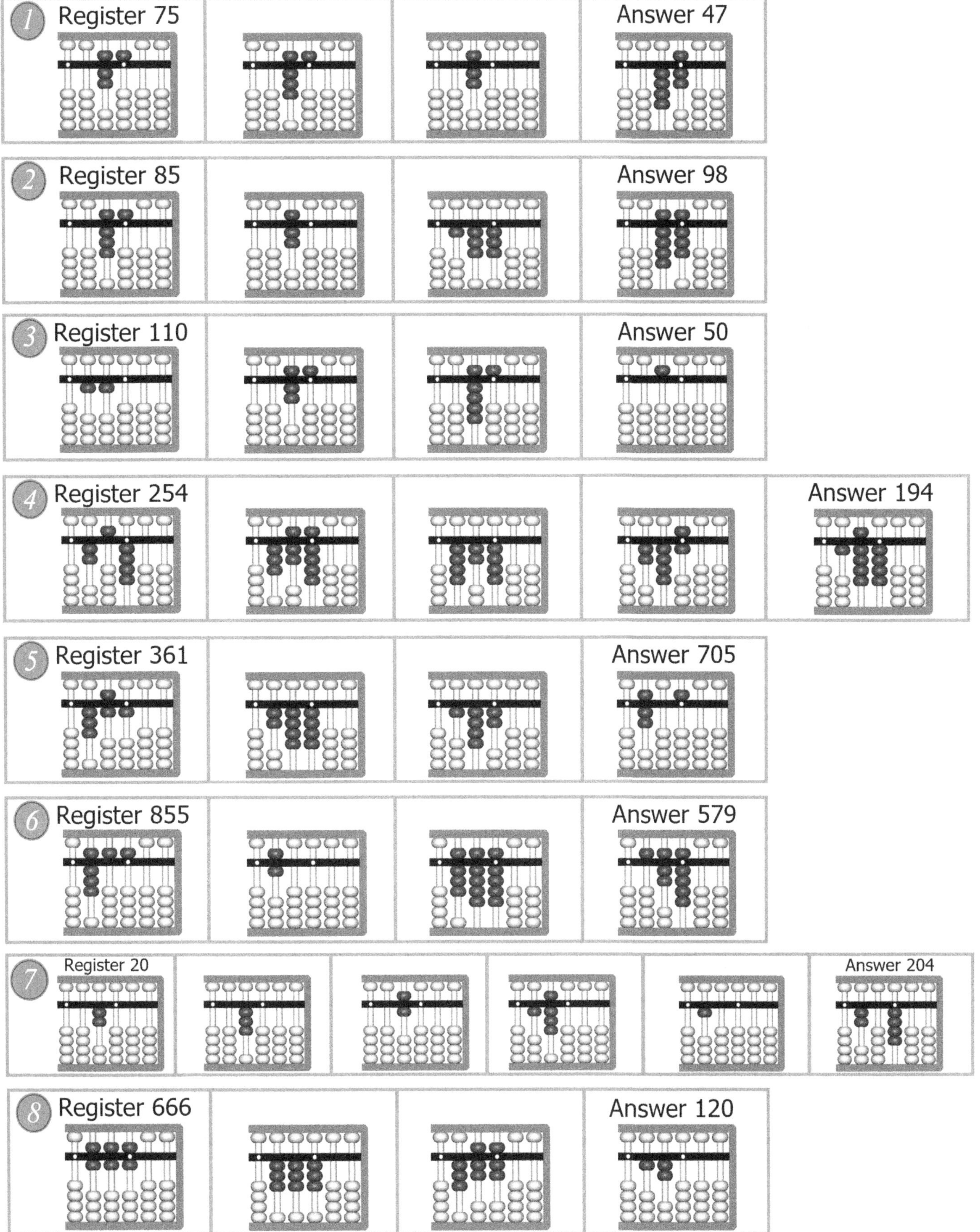
1
Register 75
Answer 47
2
Register 85
Answer 98
3
Register 110
Answer 50
4
Register 254
Answer 194
5
Register 361
Answer 705
6
Register 855
Answer 579
7
Register 20
Answer 204
8
Register 666
Answer 120

ANSWERS - 10

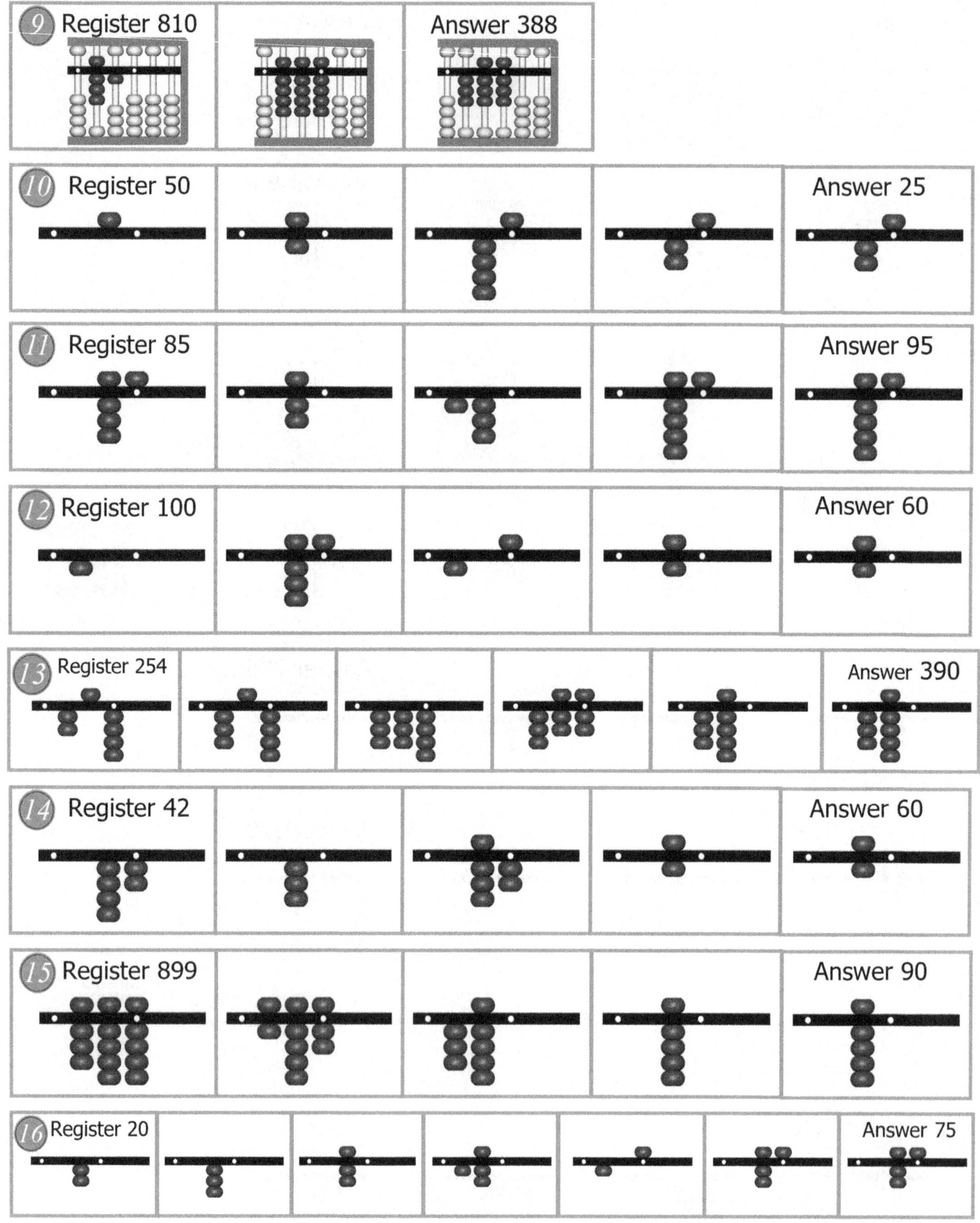

9 Register 810
Answer 388
10 Register 50
Answer 25
11 Register 85
Answer 95
12 Register 100
Answer 60
13 Register 254
Answer 390
14 Register 42
Answer 60
15 Register 899
Answer 90
16 Register 20
Answer 75

ANSWERS - Reusable - Page 1

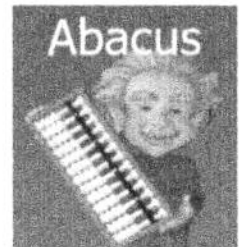

A

1	25	20	45
2	25	22	47
3	52	33	85
4	16	12	28
5	25	22	47
6	44	20	64
7	52	41	93
8	39	25	64
9	12	9	21
10	79	75	154
11	95	82	177
12	32	14	46
13	65	42	107
14	12	32	44
15	72	45	117
16	85	65	150
17	25	22	47
18	10	8	18
19	11	5	16
20	51	21	72
21	64	64	128
22	48	45	93
23	85	65	150
24	48	18	66
25	12	8	20
26	8	3	11
27	21	1	22
28	99	98	197
29	47	12	59
30	76	3	79

B

1	65	55	120
2	23	22	45
3	60	32	92
4	26	12	38
5	14	12	26
6	44	41	85
7	65	52	117
8	63	37	100
9	74	15	89
10	85	75	160
11	88	82	170
12	36	18	54
13	54	42	96
14	26	32	58
15	85	20	105
16	55	25	80
17	92	6	98
18	75	12	87
19	76	11	87
20	65	55	120
21	66	64	130
22	49	22	71
23	85	66	151
24	91	18	109
25	21	10	31
26	26	5	31
27	26	6	32
28	52	23	75
29	58	47	105
30	56	26	82

C

1	56	55	111
2	66	25	91
3	41	32	73
4	45	16	61
5	12	10	22
6	65	44	109
7	13	12	25
8	55	36	91
9	96	12	108
10	85	79	164
11	30	19	49
12	30	32	62
13	25	24	49
14	30	12	42
15	65	72	137
16	54	32	86
17	47	6	53
18	60	12	72
19	13	11	24
20	56	55	111
21	96	64	160
22	55	45	100
23	44	23	67
24	22	18	40
25	45	10	55
26	85	3	88
27	58	9	67
28	99	12	111
29	90	47	137
30	88	3	91

D

1	82	55	137
2	28	25	53
3	42	32	74
4	32	22	54
5	82	22	104
6	25	18	43
7	76	42	118
8	33	32	65
9	21	20	41
10	55	25	80
11	30	19	49
12	36	32	68
13	65	24	89
14	12	12	24
15	72	72	144
16	85	32	117
17	25	6	31
18	39	36	75
19	55	54	109
20	51	26	77
21	64	23	87
22	32	30	62
23	92	5	97
24	75	6	81
25	76	20	96
26	85	65	150
27	58	8	66
28	99	14	113
29	90	47	137
30	77	11	88

ANSWERS - Reusable - Page 2

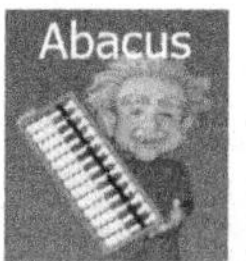

A			
1	21	55	76
2	31	25	56
3	41	32	73
4	18	22	40
5	25	36	61
6	45	18	63
7	52	66	118
8	32	32	64
9	12	12	24
10	77	25	102
11	95	21	116
12	31	32	63
13	65	25	90
14	12	13	25
15	72	77	149
16	85	32	117
17	85	6	91
18	10	66	76
19	11	54	65
20	52	26	78
21	64	23	87
22	48	32	80
23	85	9	94
24	78	6	84
25	12	22	34
26	8	55	63
27	21	8	29
28	99	15	114
29	88	47	135
30	76	12	88

B			
1	52	55	107
2	63	22	85
3	41	32	73
4	42	12	54
5	96	12	108
6	85	41	126
7	30	55	85
8	30	37	67
9	66	15	81
10	30	75	105
11	65	88	153
12	54	18	72
13	55	42	97
14	60	32	92
15	13	22	35
16	56	25	81
17	99	6	105
18	60	12	72
19	13	12	25
20	56	56	112
21	96	77	173
22	55	44	99
23	44	66	110
24	33	18	51
25	45	10	55
26	85	8	93
27	58	6	64
28	99	25	124
29	25	47	72
30	26	26	52

C			
1	56	65	121
2	66	23	89
3	41	60	101
4	45	26	71
5	12	14	26
6	65	44	109
7	13	65	78
8	55	63	118
9	96	74	170
10	85	85	170
11	30	88	118
12	30	36	66
13	25	54	79
14	30	26	56
15	65	85	150
16	54	55	109
17	47	92	139
18	60	75	135
19	13	76	89
20	56	65	121
21	96	66	162
22	55	49	104
23	44	85	129
24	22	91	113
25	45	21	66
26	85	26	111
27	58	26	84
28	99	52	151
29	90	58	148
30	76	56	132

D			
1	33	56	89
2	30	66	96
3	25	45	70
4	30	45	75
5	65	12	77
6	66	65	131
7	47	13	60
8	60	35	95
9	13	96	109
10	55	88	143
11	30	30	60
12	36	30	66
13	64	25	89
14	12	39	51
15	72	45	117
16	85	54	139
17	52	47	99
18	39	60	99
19	55	34	89
20	51	56	107
21	58	96	154
22	32	55	87
23	88	45	133
24	75	22	97
25	76	25	101
26	85	9	94
27	25	58	83
28	99	5	104
29	45	90	135
30	88	76	164

ANSWERS - Reusable - Page 3

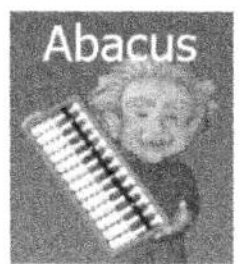

A

1	25	-20	5
2	25	-13	12
3	52	-30	22
4	85	-41	44
5	25	-9	16
6	66	-54	12
7	52	-13	39
8	39	-20	19
9	74	-20	54
10	79	-15	64
11	95	-65	30
12	32	-32	0
13	65	-7	58
14	12	-8	4
15	72	-45	27
16	85	-65	20
17	25	-22	3
18	10	-8	2
19	11	-5	6
20	51	-40	11
21	64	-33	31
22	48	-3	45
23	85	-2	83
24	48	-7	41
25	12	-5	7
26	88	-41	47
27	21	-14	7
28	99	-2	97
29	47	-12	35
30	76	-3	73

B

1	42	-3	39
2	32	-6	26
3	45	-8	37
4	65	-4	61
5	22	-3	19
6	37	-15	22
7	64	-32	32
8	51	-41	10
9	64	-50	14
10	45	-6	39
11	65	-3	62
12	64	-23	41
13	12	-10	2
14	94	-4	90
15	55	-47	8
16	55	-30	25
17	92	-53	39
18	75	-23	52
19	76	-55	21
20	65	-5	60
21	85	-70	15
22	49	-19	30
23	85	-14	71
24	91	-5	86
25	21	-6	15
26	26	-1	25
27	26	-12	14
28	52	-7	45
29	58	-47	11
30	56	-26	30

C

1	66	-33	33
2	63	-35	28
3	39	-12	27
4	45	-8	37
5	22	-12	10
6	65	-5	60
7	44	-41	3
8	55	-8	47
9	96	-15	81
10	85	-9	76
11	30	-4	26
12	30	-12	18
13	25	-16	9
14	88	-74	14
15	65	-33	32
16	98	-66	32
17	47	-6	41
18	60	-13	47
19	13	-2	11
20	56	-51	5
21	96	-12	84
22	55	-44	11
23	44	-33	11
24	22	-3	19
25	45	-2	43
26	85	-9	76
27	58	-5	53
28	99	-41	58
29	90	-14	76
30	76	-2	74

D

1	55	-10	45
2	20	-8	12
3	40	-14	26
4	32	-13	19
5	82	-9	73
6	25	-18	7
7	76	-15	61
8	33	-14	19
9	66	-44	22
10	55	-12	43
11	30	-16	14
12	36	-16	20
13	65	-45	20
14	36	-12	24
15	72	-33	39
16	85	-25	60
17	25	-12	13
18	39	-10	29
19	55	-24	31
20	51	-8	43
21	64	-12	52
22	32	-5	27
23	92	-41	51
24	75	-8	67
25	76	-15	61
26	85	-9	76
27	58	-4	54
28	99	-12	87
29	90	-23	67
30	77	-6	71

ANSWERS - Reusable - Page 4

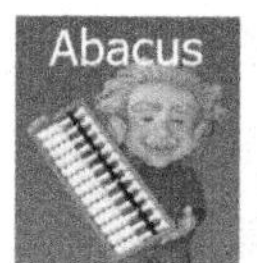

A			
1	44	-10	34
2	35	-8	27
3	45	-14	31
4	65	-13	52
5	25	-9	16
6	37	-4	33
7	64	-15	49
8	44	-14	30
9	64	-25	39
10	45	-12	33
11	66	-16	50
12	32	-16	16
13	65	-32	33
14	12	-12	0
15	72	-33	39
16	88	-25	63
17	25	-12	13
18	10	-8	2
19	55	-25	30
20	51	-8	43
21	66	-12	54
22	48	-8	40
23	85	-41	44
24	44	-8	36
25	12	-11	1
26	88	-9	79
27	32	-4	28
28	99	-14	85
29	47	-23	24
30	77	-6	71

B			
1	55	-33	22
2	55	-35	20
3	77	-12	65
4	85	-8	77
5	30	-14	16
6	44	-5	39
7	99	-88	11
8	88	-8	80
9	65	-15	50
10	78	-9	69
11	47	-7	40
12	64	-12	52
13	85	-16	69
14	94	-74	20
15	55	-21	34
16	87	-66	21
17	92	-6	86
18	75	-13	62
19	76	-7	69
20	77	-51	26
21	85	-15	70
22	21	-12	9
23	85	-33	52
24	91	-3	88
25	21	-2	19
26	55	-9	46
27	26	-8	18
28	52	-41	11
29	78	-14	64
30	56	-4	52

C			
1	66	-3	63
2	85	-13	72
3	25	-7	18
4	64	-4	60
5	55	-18	37
6	51	-15	36
7	77	-32	45
8	32	-21	11
9	92	-50	42
10	37	-6	31
11	76	-3	73
12	85	-23	62
13	58	-10	48
14	99	-9	90
15	90	-47	43
16	88	-30	58
17	96	-53	43
18	60	-5	55
19	65	-55	10
20	56	-5	51
21	77	-70	7
22	55	-35	20
23	44	-14	30
24	55	-5	50
25	45	-7	38
26	85	-15	70
27	75	-64	11
28	80	-7	73
29	90	-80	10
30	79	-26	53

D			
1	67	-20	47
2	55	-26	29
3	92	-30	62
4	75	-43	32
5	84	-9	75
6	65	-32	33
7	97	-13	84
8	33	-25	8
9	94	-20	74
10	55	-27	28
11	99	-65	34
12	36	-14	22
13	66	-8	58
14	36	-9	27
15	84	-45	39
16	85	-36	49
17	25	-22	3
18	39	-7	32
19	49	-5	44
20	51	-31	20
21	64	-33	31
22	91	-3	88
23	92	-74	18
24	83	-7	76
25	76	-4	72
26	73	-42	31
27	58	-14	44
28	99	-15	84
29	90	-75	15
30	82	-43	39

ANSWERS - Reusable - Page 5

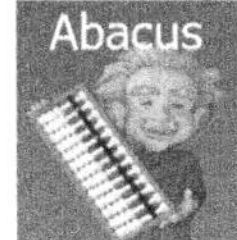

A

1	41	20	-18	43
2	16	22	-12	26
3	44	32	-15	61
4	33	16	-12	37
5	20	22	-20	22
6	45	41	-12	74
7	72	52	-32	92
8	62	36	-41	57
9	80	12	-8	84
10	14	75	-55	34
11	23	82	-12	93
12	36	14	-19	31
13	20	42	-19	43
14	66	32	-15	83
15	45	45	-62	28
16	55	65	-32	88
17	12	22	-5	29
18	41	10	-8	43
19	25	11	-6	30
20	55	51	-2	104
21	99	64	-45	118
22	32	45	-15	62
23	15	65	-15	65
24	13	18	-6	25
25	47	12	-6	53
26	88	3	-12	79
27	66	1	-15	52
28	82	98	-75	105
29	92	47	-33	106
30	24	76	-66	34

B

1	15	85	-5	95
2	23	15	-12	26
3	60	45	-15	90
4	26	35	-8	53
5	14	23	-4	33
6	44	45	-12	77
7	55	74	-12	117
8	63	62	-25	100
9	74	82	-9	147
10	46	14	-54	6
11	14	38	-13	39
12	36	33	-20	49
13	14	20	-20	14
14	26	65	-16	75
15	85	45	-65	65
16	55	54	-32	77
17	92	12	-7	97
18	75	12	-8	79
19	76	13	-3	86
20	65	55	-2	118
21	32	96	-45	83
22	26	32	-14	44
23	85	12	-33	64
24	91	13	-4	100
25	21	47	-2	66
26	26	85	-9	102
27	26	66	-5	87
28	4	82	-42	44
29	18	91	-14	95
30	9	23	-2	30

C

1	20	55	-4	71
2	25	22	-3	44
3	32	32	-6	58
4	16	18	-6	28
5	22	22	-5	39
6	44	41	-2	83
7	52	52	-6	98
8	36	37	-10	63
9	12	12	-9	15
10	79	75	-22	132
11	19	82	-2	99
12	32	18	-12	38
13	24	42	-6	60
14	12	32	-4	40
15	72	20	-3	89
16	32	65	-10	87
17	64	6	-24	46
18	10	11	-13	8
19	11	11	-2	20
20	51	55	-51	55
21	64	64	-12	116
22	48	45	-44	49
23	65	66	-33	98
24	18	18	-3	33
25	12	10	-2	20
26	32	3	-9	26
27	35	1	-5	31
28	98	99	-41	156
29	47	47	-14	80
30	76	76	-2	150

ANSWERS - Reusable - Page 6

A

1	52	74	-18	108
2	16	52	-12	56
3	33	82	-15	100
4	33	16	-12	37
5	20	38	-13	45
6	62	52	-12	102
7	72	21	-32	61
8	62	65	-32	95
9	80	40	-8	112
10	16	54	-55	15
11	23	12	-9	26
12	36	8	-19	25
13	44	13	-19	38
14	66	52	-15	103
15	45	96	-52	89
16	55	12	-32	35
17	44	8	-5	47
18	41	6	-8	39
19	25	47	-6	66
20	55	85	-2	138
21	99	66	-33	132
22	32	20	-15	37
23	15	91	-15	91
24	13	16	-6	23
25	47	12	-4	55
26	99	3	-12	90
27	66	3	-15	54
28	82	98	-13	167
29	92	47	-33	106
30	24	20	-23	21

B

1	22	85	-5	102
2	23	82	-12	93
3	60	12	-16	56
4	16	12	-8	20
5	22	47	-4	65
6	20	85	-9	96
7	52	16	-8	60
8	36	88	-25	99
9	12	91	-9	94
10	79	23	-44	58
11	19	17	-13	23
12	32	9	-20	21
13	24	9	-18	15
14	12	28	-12	28
15	16	47	-15	48
16	32	76	-5	103
17	64	22	-20	66
18	10	55	-12	53
19	80	13	-32	61
20	14	30	-41	3
21	23	96	-6	113
22	36	32	-55	13
23	20	11	-12	19
24	66	11	-4	73
25	45	47	-16	76
26	55	55	-9	101
27	77	66	-25	118
28	4	82	-42	44
29	18	36	-14	40
30	9	39	-2	46

C

1	10	55	-4	61
2	10	22	-3	29
3	32	36	-6	62
4	16	18	-3	31
5	22	22	-20	24
6	44	20	-12	52
7	64	50	-15	99
8	20	37	-12	45
9	11	12	-20	3
10	51	68	-12	107
11	64	55	-32	87
12	95	18	-41	72
13	85	42	-8	119
14	91	13	-55	49
15	21	5	-3	23
16	26	60	-10	76
17	28	6	-24	10
18	4	20	-13	11
19	11	41	-2	50
20	60	55	-51	64
21	64	64	-12	116
22	48	8	-44	12
23	65	76	-33	108
24	18	55	-3	70
25	99	10	-2	107
26	32	6	-9	29
27	35	12	-5	42
28	98	41	-41	98
29	47	20	-14	53
30	76	88	-2	162

ANSWERS - Reusable - Page 7

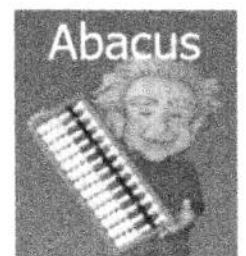

A

1	88	74	-3	159
2	15	52	-3	64
3	48	82	-6	124
4	35	16	-7	44
5	25	38	-5	58
6	45	55	-2	98
7	75	21	-6	90
8	62	25	-12	75
9	82	40	-9	113
10	14	54	-22	46
11	39	12	-15	36
12	33	8	-12	29
13	20	15	-6	29
14	65	52	-10	107
15	47	96	-3	140
16	54	12	-10	56
17	18	9	-24	3
18	12	6	-13	5
19	46	47	-2	91
20	55	85	-51	89
21	96	67	-12	151
22	32	20	-44	8
23	12	91	-33	70
24	13	16	-14	15
25	88	12	-2	98
26	85	3	-20	68
27	66	12	-5	73
28	82	98	-44	136
29	85	47	-14	118
30	23	26	-12	37

B

1	66	85	-15	136
2	85	88	-12	161
3	25	12	-15	22
4	65	12	-18	59
5	55	47	-20	82
6	51	95	-13	133
7	85	16	-20	81
8	32	88	-41	79
9	92	99	-7	184
10	37	23	-45	15
11	88	17	-13	92
12	85	15	-15	85
13	58	9	-22	45
14	99	36	-15	120
15	90	47	-65	72
16	99	76	-13	162
17	96	22	-7	111
18	60	56	-6	110
19	65	15	-9	71
20	56	30	-4	82
21	87	96	-26	157
22	55	32	-12	75
23	44	15	-15	44
24	55	11	-7	59
25	46	47	-4	89
26	85	60	-12	133
27	85	66	-16	135
28	80	82	-44	118
29	90	37	-33	94
30	79	42	-55	66

C

1	55	55	-6	104
2	30	25	-8	47
3	25	36	-7	54
4	33	18	-4	47
5	65	40	-12	93
6	66	20	-12	74
7	47	55	-22	80
8	60	37	-42	55
9	45	12	-30	27
10	55	72	-6	121
11	30	55	-3	82
12	36	18	-33	21
13	64	42	-10	96
14	24	13	-6	31
15	72	5	-35	42
16	85	60	-30	115
17	55	6	-42	19
18	39	25	-23	41
19	55	41	-55	41
20	51	55	-42	64
21	65	64	-12	117
22	32	28	-5	55
23	88	76	-72	92
24	85	55	-18	122
25	76	12	-12	76
26	85	6	-15	76
27	25	12	-9	28
28	88	52	-9	131
29	45	20	-7	58
30	75	88	-8	155

ANSWERS - Reusable - Page 8

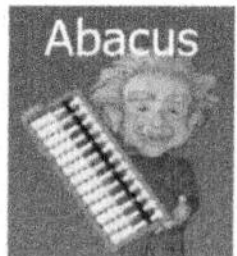

A

1	10	20	-15	-5	10
2	30	22	-12	-3	37
3	40	32	-15	-2	55
4	5	16	-14	-4	3
5	10	22	-20	-5	7
6	30	41	-13	-1	57
7	10	52	-30	-5	27
8	30	36	-41	-10	15
9	10	12	-9	-9	4
10	30	78	-54	-20	34
11	10	19	-13	-2	14
12	30	32	-20	-12	30
13	10	24	-20	-5	9
14	30	6	-15	-4	17
15	10	72	-65	-3	14
16	30	32	-32	-17	13
17	42	6	-7	-24	17
18	30	10	-8	-13	19
19	10	11	-3	-4	14
20	30	51	-2	-51	28
21	10	64	-45	-12	17
22	30	45	-12	-40	23
23	10	65	-15	-33	27
24	30	18	-6	-3	39
25	10	12	-4	-2	16
26	30	3	-12	-7	14
27	47	1	-15	-5	28
28	30	98	-74	-41	13
29	10	47	-33	-14	10
30	30	76	-66	-2	38

B

1	15	-3	63	-5	70
2	22	-6	15	-3	28
3	60	-8	42	-2	92
4	26	-4	35	-4	53
5	12	-3	27	-5	31
6	44	-12	45	-1	76
7	52	-32	65	-5	80
8	63	-41	75	-10	87
9	74	-50	82	-9	97
10	21	-5	14	-20	10
11	12	-3	38	-2	45
12	36	-23	32	-12	33
13	14	-10	20	-5	19
14	25	-7	65	-4	79
15	85	-47	41	-3	76
16	52	-30	54	-17	59
17	92	-53	12	-24	27
18	74	-23	9	-13	47
19	76	-55	13	-4	30
20	65	-42	54	-51	26
21	32	-14	96	-12	102
22	25	-6	32	-40	11
23	85	-70	9	-6	18
24	91	-19	2	-3	71
25	21	-14	47	-2	52
26	25	-3	85	-7	100
27	26	-6	61	-5	76
28	2	-1	82	-41	42
29	18	-14	90	-14	80
30	9	-7	23	-2	23

C

1	20	20	-10	-5	35
2	65	22	-8	-12	75
3	41	32	-13	-15	58
4	54	16	-14	-6	50
5	12	22	-7	-4	23
6	9	41	-8	-12	30
7	13	52	-3	-15	47
8	54	36	-2	-74	14
9	96	12	-45	-9	54
10	30	75	-12	-54	39
11	10	82	-15	-13	64
12	30	14	-6	-20	18
13	10	42	-4	-20	28
14	30	32	-12	-15	35
15	65	20	-32	-45	8
16	54	65	-35	-32	52
17	45	6	-12	-7	32
18	60	10	-7	-8	55
19	13	11	-12	-3	9
20	54	51	-5	-2	98
21	96	64	-41	-45	74
22	32	45	-8	-12	57
23	9	65	-15	-33	26
24	2	18	-6	-3	11
25	47	12	-4	-2	53
26	85	3	-12	-7	69
27	61	1	-15	-5	42
28	82	98	-74	-41	65
29	90	47	-33	-14	90
30	23	76	-66	-2	31

ANSWERS - Reusable - Page 9

A

1	85	20	-15	-5	85
2	15	22	-12	-3	22
3	45	32	-15	-2	60
4	35	16	-14	-4	33
5	23	22	-20	-5	20
6	45	41	-13	-1	72
7	74	52	-30	-5	91
8	62	36	-41	-10	47
9	82	12	-9	-9	76
10	14	75	-54	-20	15
11	38	82	-13	-2	105
12	33	14	-20	-12	15
13	20	42	-20	-5	37
14	65	32	-15	-4	78
15	45	45	-65	-3	22
16	54	65	-32	-17	70
17	12	22	-7	-24	3
18	12	10	-8	-13	1
19	13	11	-3	-4	17
20	55	51	-2	-51	53
21	96	64	-45	-12	103
22	32	45	-12	-40	25
23	12	65	-15	-33	29
24	13	18	-6	-3	22
25	47	12	-4	-2	53
26	85	3	-12	-7	69
27	66	1	-15	-5	47
28	82	98	-74	-41	65
29	91	47	-33	-14	91
30	23	76	-66	-2	31

B

1	15	-3	20	-5	27
2	23	-6	25	-12	30
3	60	-8	32	-15	69
4	26	-4	16	-8	30
5	14	-3	22	-4	29
6	44	-15	44	-12	61
7	55	-32	52	-12	63
8	63	-41	36	-25	33
9	74	-50	12	-9	27
10	21	-6	79	-54	40
11	14	-3	19	-13	17
12	36	-23	32	-20	25
13	14	-10	24	-20	8
14	26	-4	12	-16	18
15	85	-47	72	-65	45
16	55	-30	32	-32	25
17	92	-53	6	-7	38
18	75	-23	10	-8	54
19	76	-55	11	-3	29
20	65	-42	51	-2	72
21	32	-14	64	-45	37
22	26	-5	48	-14	55
23	85	-70	65	-33	47
24	91	-19	18	-4	86
25	21	-14	12	-2	17
26	26	-5	3	-9	15
27	26	-6	1	-5	16
28	4	-1	98	-42	59
29	18	-12	47	-14	39
30	9	-7	76	-2	76

C

1	21	55	-10	-4	72
2	66	22	-8	-3	85
3	41	32	-14	-6	67
4	45	18	-14	-6	43
5	12	22	-9	-5	20
6	9	41	-8	-2	40
7	13	52	-3	-6	56
8	55	37	-2	-10	80
9	96	12	-44	-9	55
10	31	75	-12	-22	72
11	10	82	-16	-2	74
12	30	18	-6	-12	30
13	12	42	-4	-6	44
14	30	32	-12	-4	46
15	65	20	-33	-3	49
16	54	65	-35	-10	74
17	47	6	-12	-24	17
18	60	11	-8	-13	50
19	13	11	-12	-2	10
20	54	55	-5	-51	53
21	96	64	-41	-12	107
22	32	45	-8	-44	25
23	11	66	-15	-33	29
24	2	18	-9	-3	8
25	45	10	-4	-2	49
26	85	3	-12	-9	67
27	58	1	-16	-5	38
28	82	99	-74	-41	66
29	90	47	-33	-14	90
30	25	76	-66	-2	33

ANSWERS - Reusable - Page 10

A

1	30	20	-15	-4	31
2	65	22	-15	-3	69
3	41	25	-15	-5	46
4	54	16	-23	-4	43
5	14	22	-20	-5	11
6	9	44	-14	-1	38
7	16	52	-30	-7	31
8	54	74	-39	-10	79
9	99	12	-9	-9	93
10	30	77	-54	-20	33
11	12	82	-13	-12	69
12	30	15	-21	-12	12
13	10	42	-20	-5	27
14	66	32	-12	-4	82
15	65	52	-65	-3	49
16	54	65	-25	-17	77
17	52	22	-7	-24	43
18	60	12	-9	-13	50
19	13	11	-3	-4	17
20	54	44	-6	-51	41
21	78	64	-45	-15	82
22	60	32	-15	-40	37
23	9	65	-15	-33	26
24	2	25	-6	-3	18
25	52	12	-4	-2	58
26	85	3	-12	-11	65
27	61	1	-15	-5	42
28	82	88	-74	-41	55
29	99	47	-33	-16	97
30	23	80	-66	-3	34

B

1	40	-5	40	-5	70
2	22	-15	25	-10	22
3	32	-32	25	-12	13
4	16	-8	16	-8	16
5	41	-5	22	-4	54
6	55	-25	26	-12	44
7	52	-40	53	-14	51
8	36	-23	33	-25	21
9	12	-11	15	-9	7
10	75	-45	70	-54	46
11	87	-11	20	-13	83
12	63	-27	33	-20	49
13	42	-20	13	-20	15
14	32	-18	54	-16	52
15	84	-74	72	-66	16
16	65	-25	60	-32	68
17	16	-8	20	-7	21
18	10	-8	15	-8	9
19	11	-5	12	-3	15
20	66	-2	12	-2	74
21	64	-20	64	-45	63
22	45	-14	24	-14	41
23	65	-25	66	-33	73
24	18	-9	18	-4	23
25	24	-2	21	-2	41
26	35	-10	9	-9	25
27	36	-6	3	-5	28
28	98	-3	98	-42	151
29	47	-14	54	-14	73
30	45	-7	25	-2	61

C

1	36	55	-4	-15	72
2	25	32	-3	-20	34
3	56	32	-6	-15	67
4	16	19	-6	-13	16
5	62	22	-20	-20	44
6	25	42	-12	-13	42
7	52	52	-17	-30	57
8	33	52	-12	-32	41
9	12	35	-20	-9	18
10	74	75	-10	-54	85
11	19	88	-32	-13	62
12	52	18	-42	-20	8
13	32	42	-8	-20	46
14	54	32	-45	-15	26
15	72	20	-3	-12	77
16	60	55	-10	-32	73
17	18	18	-24	-7	5
18	15	11	-13	-8	5
19	12	11	-2	-3	18
20	12	65	-51	-2	24
21	64	25	-12	-45	32
22	48	45	-40	-12	41
23	66	66	-33	-9	90
24	18	27	-3	-6	36
25	88	10	-2	-4	92
26	18	3	-5	-12	4
27	20	8	-5	-15	8
28	98	99	-32	-74	91
29	36	47	-14	-51	18
30	35	25	-2	-33	25

ANSWERS - Reusable - Page 11

A

1	88	20	-12	-2	-5	89
2	17	22	-12	-4	-3	20
3	44	33	-15	-4	-2	56
4	30	16	-16	-4	-4	22
5	30	22	-20	-9	-5	18
6	44	44	-13	-1	-1	73
7	77	52	-20	-5	-5	99
8	63	26	-41	-12	-10	26
9	82	12	-9	-12	-9	64
10	15	78	-45	-22	-20	6
11	33	82	-13	-3	-2	97
12	35	15	-15	-12	-12	11
13	25	42	-20	-5	-5	37
14	66	33	-15	-4	-4	76
15	44	45	-65	-6	-3	15
16	50	66	-16	-17	-17	66
17	40	22	-9	-24	-24	5
18	30	12	-9	-13	-13	7
19	78	11	-9	-4	-4	72
20	90	51	-12	-25	-51	53
21	95	64	-25	-12	-12	110
22	67	47	-12	-30	-40	32
23	46	65	-15	-33	-33	30
24	28	78	-7	-9	-3	87
25	82	12	-4	-2	-2	86
26	65	36	-12	-17	-7	65
27	46	6	-16	-5	-5	26
28	95	14	-44	-21	-41	3
29	52	55	-33	-33	-14	27
30	32	85	-55	-6	-2	54

B

1	15	-6	40	-5	14	58
2	22	-6	25	-12	25	54
3	60	-7	33	-32	30	84
4	26	-4	16	-8	16	46
5	15	-12	22	-4	36	57
6	44	-15	25	-12	44	86
7	56	-22	52	-40	52	98
8	63	-41	33	-25	36	66
9	74	-30	12	-9	12	59
10	22	-6	70	-54	79	111
11	14	-3	19	-11	19	38
12	35	-23	33	-20	33	58
13	15	-10	24	-20	24	33
14	26	-6	54	-16	12	70
15	88	-47	72	-70	62	105
16	55	-30	60	-35	33	83
17	99	-42	6	-8	6	61
18	75	-23	12	-8	14	70
19	85	-50	12	-3	11	55
20	65	-42	12	-2	52	85
21	32	-12	64	-60	66	90
22	13	-5	48	-14	44	86
23	85	-70	66	-30	65	116
24	91	-18	18	-4	44	131
25	20	-12	22	-2	12	40
26	26	-9	9	-10	8	24
27	36	-9	3	-5	1	26
28	14	-9	98	-25	44	122
29	18	-6	25	-14	25	48
30	10	-8	25	-6	12	33

ANSWERS - Reusable - Page 12

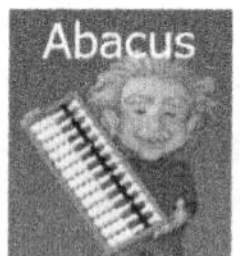

A

1	20	20	-4	-2	20	54
2	65	22	-3	-4	22	102
3	41	33	-6	-4	32	96
4	54	16	-3	-4	16	79
5	12	22	-20	-9	22	27
6	9	44	-12	-1	41	81
7	13	52	-15	-5	52	97
8	54	26	-12	-12	36	92
9	96	12	-20	-12	12	88
10	30	78	-12	-22	75	149
11	10	82	-32	-3	82	139
12	30	33	-41	-12	14	24
13	10	42	-8	-5	42	81
14	30	33	-55	-4	32	36
15	65	45	-3	-6	20	121
16	54	66	-10	-17	65	158
17	45	22	-24	-24	6	25
18	60	12	-13	-13	10	56
19	13	11	-2	-4	11	29
20	54	51	-51	-25	51	80
21	96	64	-12	-12	64	200
22	60	47	-44	-30	45	78
23	9	65	-33	-33	65	73
24	2	78	-3	-9	18	86
25	47	12	-2	-2	12	67
26	85	36	-9	-17	3	98
27	61	6	-5	-5	1	58
28	82	14	-41	-21	98	132
29	90	55	-14	-33	47	145
30	23	85	-2	-6	76	176

B

1	41	-15	40	-5	14	75
2	16	-12	25	-12	25	42
3	44	-15	33	-32	30	60
4	33	-14	16	-8	16	43
5	20	-20	22	-4	36	54
6	45	-13	25	-26	44	75
7	72	-30	52	-40	52	106
8	62	-41	33	-25	36	65
9	80	-9	12	-9	12	86
10	64	-54	70	-54	79	105
11	23	-13	19	-11	19	37
12	36	-20	33	-38	33	44
13	36	-20	22	-20	24	42
14	66	-15	54	-16	12	101
15	45	-3	72	-70	62	106
16	55	-32	60	-35	33	81
17	85	-7	6	-8	6	82
18	41	-8	12	-8	14	51
19	25	-3	12	-3	11	42
20	55	-2	12	-2	52	115
21	99	-45	64	-60	66	124
22	32	-12	48	-14	44	98
23	16	-15	66	-30	65	102
24	13	-6	18	-4	44	65
25	47	-4	22	-2	12	75
26	88	-12	9	-10	8	83
27	66	-15	3	-5	1	50
28	82	-74	98	-25	44	125
29	92	-33	25	-14	25	95
30	99	-66	25	-6	12	64

ANSWERS - Reusable - Page 13

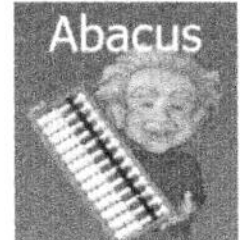

A

1	102	242	-3	-200	20	161
2	22	12	-6	-4	22	46
3	123	62	-8	-4	32	205
4	18	121	-100	-4	16	51
5	22	6	-3	-9	22	38
6	441	14	-15	-210	41	271
7	225	11	-32	-5	52	251
8	550	111	-41	-221	36	435
9	12	66	-50	-12	12	28
10	321	44	-6	-22	75	412
11	82	65	-3	-3	82	223
12	18	623	-23	-12	14	620
13	42	42	-10	-22	42	94
14	251	33	-4	-4	32	308
15	20	45	-47	-6	20	32
16	65	412	-30	-17	65	495
17	333	22	-53	-120	6	188
18	412	12	-23	-13	10	398
19	11	223	-55	-4	11	186
20	550	51	-400	-25	51	227
21	213	821	-70	-300	64	728
22	45	90	-19	-60	45	101
23	166	65	-121	-33	65	142
24	18	78	-5	-9	18	100
25	362	12	-6	-2	12	378
26	875	36	-418	-17	3	479
27	365	6	-12	-5	1	355
28	999	14	-555	-21	98	535
29	400	55	-47	-33	47	422
30	76	288	-200	-6	76	234

B

1	300	-15	40	-10	14	329
2	22	-12	520	-8	25	547
3	152	-140	33	-14	30	61
4	36	-14	160	-13	16	185
5	220	-20	22	-9	36	249
6	75	-13	125	-18	44	213
7	882	-255	52	-15	52	716
8	140	-41	33	-44	36	124
9	425	-250	12	-44	12	155
10	632	-155	70	-120	79	506
11	120	-13	19	-16	19	129
12	65	-20	33	-16	33	95
13	600	-200	22	-310	24	136
14	100	-15	54	-12	12	139
15	110	-35	72	-33	62	176
16	665	-32	60	-150	33	576
17	452	-200	6	-120	6	144
18	210	-8	12	-10	14	218
19	110	-3	130	-24	11	224
20	230	-25	12	-8	52	261
21	64	-45	64	-12	66	137
22	125	-12	233	-155	44	235
23	610	-250	66	-200	65	291
24	180	-80	18	-8	44	154
25	120	-4	22	-15	12	135
26	125	-25	9	-9	8	108
27	225	-106	3	-40	1	83
28	99	-25	188	-12	44	294
29	850	-450	25	-23	25	427
30	320	-120	25	-6	12	231

ANSWERS - Reusable - Page 14

A

1	120	242	-3	-12	20	367
2	222	100	-120	-4	22	220
3	133	62	-8	-25	32	194
4	16	121	-100	-4	16	49
5	250	6	-120	-9	22	149
6	44	320	-15	-120	41	270
7	520	225	-365	-5	52	427
8	26	111	-41	-12	36	120
9	120	66	-50	-120	12	28
10	780	44	-400	-22	75	477
11	82	665	-3	-3	82	823
12	330	623	-23	-12	14	932
13	42	250	-25	-5	42	304
14	124	33	-4	-24	32	161
15	450	45	-200	-6	20	309
16	66	412	-30	-17	65	496
17	222	22	-30	-24	6	196
18	12	444	-23	-44	10	399
19	110	223	-55	-4	11	285
20	51	515	-400	-25	51	192
21	132	821	-70	-520	64	427
22	165	90	-19	-30	45	251
23	65	241	-121	-14	65	236
24	136	78	-5	-9	18	218
25	12	444	-6	-2	12	460
26	950	555	-418	-600	3	490
27	963	6	-12	-540	1	418
28	652	100	-555	-21	98	274
29	55	200	-47	-33	47	222
30	425	288	-200	-6	76	583

B

1	240	-6	120	-10	14	358
2	125	-6	420	-8	25	556
3	130	-7	55	-14	30	194
4	160	-4	160	-13	16	319
5	360	-12	140	-145	36	379
6	440	-15	125	-250	44	344
7	225	-200	52	-15	52	114
8	635	-410	33	-44	36	250
9	127	-30	12	-44	12	77
10	790	-650	70	-120	79	169
11	250	-54	19	-16	19	218
12	330	-23	33	-16	33	357
13	240	-110	200	-310	24	44
14	120	-60	54	-12	12	114
15	665	-470	72	-33	62	296
16	352	-300	500	-150	33	435
17	362	-320	600	-120	6	528
18	145	-23	12	-10	14	138
19	110	-50	130	-24	11	177
20	520	-420	12	-8	52	156
21	660	-500	64	-12	66	278
22	415	-350	233	-155	44	187
23	254	-145	66	-20	65	220
24	400	-200	18	-8	44	254
25	120	-25	22	-15	12	114
26	880	-741	9	-9	8	147
27	199	-99	3	-40	1	64
28	420	-145	188	-12	44	495
29	250	-125	25	-23	25	152
30	150	-50	25	-6	12	131

Notes

Made in the USA
Monee, IL
25 November 2019